How to Use This Workbook

This workbook has been written and developed to be used alongside the Lonsdale revision guide, OCR Gateway GCSE Science Essentials, to help you get the most out of your revision. You can use it if you are studying OCR Gateway GCSE Science at Foundation or Higher Tier.

It contains 'quick fire' questions, including multiple choice questions, matching pair exercises and short answer questions, to test your understanding of the topics covered in the revision guide.

Start by reading through a topic in the revision guide. Make notes, jotting down anything that you think will help you to remember the information. When you have finished, take a short break and then read through your notes. You might even want to try covering them up and writing them out again from memory.

Finally, work through the relevant questions in this workbook without looking at the guide or your notes.

The page headers and sub-headings in this workbook correspond with those in your revision guide, so that you can easily identify the questions that relate to each topic.

Completing the questions will help to reinforce your understanding of the topics covered in the guide and highlight any areas that need further revision.

The answers to the questions in this workbook are available in a separate booklet. There is a box at the end of each page for you to record your score. Don't worry if you get some questions wrong the first time. Just re-read the information in your revision guide and try again.

The tick boxes on the contents page let you track your revision progress: simply put a tick in the box next to each topic when you are confident that you understand it.

Good luck with your exams!

Drugs and You

Drugs

1 Match effects **A**, **B**, **C** and **D**, with the types of drug **1–4** in the table below. Enter the appropriate number in the boxes provided.

	Category of drug
1	Stimulant
2	Sedative
3	Painkiller
4	Hallucinogen

A Makes you see or hear things that are not there

B Slows down the brain and makes you feel sleepy

C Causes increased brain activity and alertness

D Reduces feelings of pain

HT

2 Fill in the missing words to complete the sentences below:

a) Stimulants ______________ the amount of transmitter substance released at synapses.

b) Depressants ______________ the amount of transmitter substance released at synapses.

Drug Classification

3 Illegal drugs are classified into three main categories under the Misuse of Drugs Act. Write **A**, **B** or **C** alongside each statement to show which category it applies to.

a) Has the heaviest prison sentences and fines for possession.

b) Includes the most dangerous drugs like heroine and cocaine.

c) Carries the lowest penalties for possession, but is still illegal.

d) Includes drugs like cannabis and anabolic steroids.

e) Includes amphetamines like speed and barbiturates.

Addiction and Rehabilitation

4 Addicts' bodies become used to drugs, and they start to need more and more to have the same effect. What is this called? Tick the correct option.

A Addiction **B** Tolerance

C Withdrawal **D** Rehabilitation

5 What are two types of withdrawal symptom?

a) ______________ **b)** ______________

Drugs and You

Alcohol

1 Alcohol is a commonly used legal drug. What type of drug is alcohol? Tick the correct option.

A Sedative **B** Analgesic

C Stimulant **D** Hallucinogen

2 **a)** Which of the following are **not** common short-term effects of drinking too much alcohol? Tick the three correct options.

A Slurred speech **B** Lack of balance

C Blurred vision **D** Liver damage

E Brain damage **F** Damage to unborn child

b) Fill in the missing word to complete the following sentence:

In the long-term, excessive alcohol consumption can damage the liver causing a disease called ______________.

Tobacco

3 **a)** Despite the risks, many people continue to smoke cigarettes. Tobacco smoke can be very dangerous. Which of these chemicals is found in smoke? Tick the three correct options.

A Tar **B** Carbon monoxide

C Nicotine **D** Cilia

E Ketamine

b) Which of these illnesses is **not** directly associated with smoking tobacco? Tick the three correct options.

A High blood pressure **B** Emphysema

C Cirrhosis **D** Bronchitis

E Arteriosclerosis **F** Paranoia

G Pneumonia

HT

4 Which two components of cigarette smoke can cause cancer?

a) ______________ **b)** ______________

Contents

Module B1: Understanding Ourselves

Module C1: Carbon Chemistry

Module P1: Energy for the Home

Contents

Module B2: Understanding our Environment

Module C2: Rocks and Metals

Module P2: Living for the Future

Fit for Life

The Circulatory System and Blood Pressure

1 Fill in the missing words to complete the sentence:

a) The circulatory system carries .... and .... in your blood to all your body's cells.

b) Energy is then released through .... respiration.

c) Circle the correct options in the following sentences:

i) The heart is a muscular **opening / pump / gap / cell** which relaxes and contracts.

ii) Two measurements are used to represent blood pressure. The first value is the **systolic / diastolic / systematic / systemic** pressure, when the blood's being forced into the arteries.

iii) The second value is the **systolic / diastolic / systematic / systemic** pressure, when the heart is **stopped / dead / tense / relaxed**.

d) Complete the following sentences:

i) The unit for measuring blood pressure is ....

ii) Normal blood pressure is ....

2 Which of the following factors may lead to high blood pressure? Tick the correct options.

A Excess weight ☐

B High stress levels ☐

C Regular exercise ☐

D Smoking ☐

E Excess alcohol ☐

F A high fat diet ☐

G Cold hands ☐

H A low fat diet ☐

HT

3 High blood pressure might cause blood vessels to burst. What is it called when vessels in the brain burst?

....

4 **a)** What does low blood pressure deprive the body of and why?

....

b) Name two symptoms of low blood pressure.

....

Pulse and Effect of Exercise

1 The graph shows Isaac's heart rate after a long distance race.

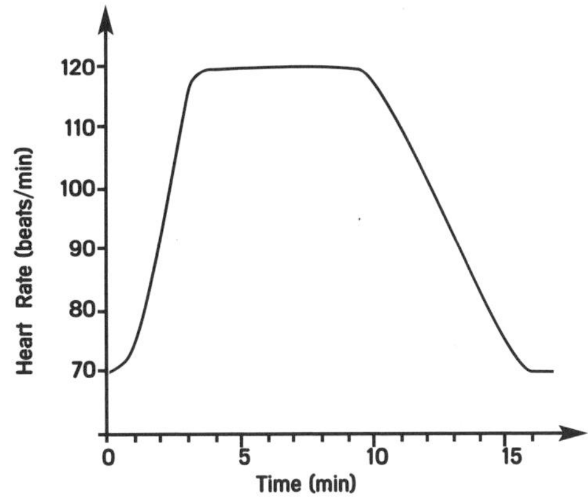

a) i) What was Isaac's resting heart rate (in beats per minute)?

....

ii) Is this rate normal?

....

b) How long did it take Isaac's pulse to return to normal? Tick the correct option.

A 15 minutes ☐ **B** 5 minutes ☐

C 120 minutes ☐ **D** 70 minutes ☐

c) Give two reasons why breathing and heart rate increase during exercise.

....

....

Health and Fitness

2 Fill in the missing words to complete the following sentences:

a) Being .... means being free from infection.

b) Being .... relates to how well your body copes during and after exercise.

HT

3 Name three different aspects of fitness.

a) .... **b)** .... **c)** ....

4 Circle the correct option in the following sentence:

Cardiovascular **energy / exercise / efficiency / effort** is often used as a measure of fitness. It refers to how well the **legs / heart / muscles / brain** copes with aerobic exercise and how quickly it **survives / beats / pumps / recovers**.

Fit for Life

Aerobic Respiration

1 Respiration is an essential process. Which of these statements about respiration are true? Tick the three correct options.

- **A** Aerobic respiration takes place in cells. ☐
- **B** Respiration releases energy. ☐
- **C** Glucose reacts with oxygen during aerobic respiration. ☐
- **D** Respiration is the scientific word for breathing. ☐

2 Which of the equations below summarises aerobic respiration? Tick the correct option.

- **A** Glucose + water → carbon dioxide + oxygen + energy ☐
- **B** Glucose + oxygen → carbon dioxide + water + energy ☐
- **C** Carbon dioxide + water → glucose + oxygen + energy ☐
- **D** Carbon dioxide + oxygen → glucose + water + energy ☐

HT

3 One of the substances needed for respiration to take place is $C_6H_{12}O_6$. What substance is this the chemical formula for?

....

Anaerobic Respiration

4 Circle the correct option in the following sentences:

a) Anaerobic respiration takes place when cells cannot get enough **water / carbon / oxygen / muscles**.

b) Energy is released and **lactic acid / ethanoic acid / water / oxygen** is produced.

c) This is a waste product and when it builds up it can cause **cramps / death / stroke / exercise**.

HT

5 During anaerobic exercise, an oxygen debt builds up. How is this debt repaid?

....

....

What's for Lunch?

A Balanced Diet

1 Fill in the missing words to complete the following sentences:

a) A balanced diet must contain .... , ... and protein.

b) Proteins are used for the .... and .... of

....

2 What two types of food are mainly used to provide energy?

a) .... **b)** ....

3 Name two other substances needed in your diet to keep your body healthy.

a) .... **b)** ....

HT

4 Give two reasons why some people alter their diet.

a) .... **b)** ....

How Much Energy is Needed?

5 Circle the correct options in the following sentence:

If a person **consumes / expends / loses / excretes** more energy than they **consume / need / lose / excrete** they may eventually become **anorexic / balanced / protein / obese.**

6 **a)** One method of checking for obesity is to divide mass (kg) by height squared (m^2).
What is this measure called? Tick the correct option.

A BMO ☐ **B** BFI ☐

C BMI ☐ **D** BIM ☐

b) What does the abbreviation above stand for? ....

Protein

7 What is the difference between an essential amino acid and a non-essential amino acid?

....

What's For Lunch?

HT Eating Disorders

1 a) Some people with eating disorders restrict what they eat and consume too little energy. What is this disorder called?

....

b) Some people with eating disorders make themselves sick after eating or take laxatives. What is this disorder called?

....

The Human Digestive System

2 Which two parts of the body are involved in physical digestion?

a) .... b) ....

3 Circle the correct options in the following sentences:

Chemical digestion uses **enzymes / diffusion / the mouth / pressure** to break down large **insoluble / soluble / obese / amino** molecules into smaller **insoluble / soluble / obese / amino** ones. These can then **diffuse / leak / osmose / disperse**.

4 During digestion, foods are broken down. Fill in the missing words to complete the sentences below:

a) .... are broken down into soluble sugars.

b) .... are broken down into amino acids.

c) .... are broken down into fatty acids and glycerol.

5 Fill in the missing words to complete the sentences below:

Hydrochloric acid is made by cells in the .... The acid kills ....

and provides the optimum pH for the .... in the stomach to work.

HT

6 Digestion allows us to get useful products from our food. How does bile help to digest fat?

....

....

Non-Infectious Diseases

1 Name three factors that can lead to non-infectious diseases.

a) .... b) .... c) ....

Cancer

2 Cancer is a non-infectious disease caused by mutations in living cells.
Which of the actions below could help to reduce the risk of cancer? Tick the two correct options.

A Smoking ☐ B Not drinking excess alcohol ☐

C Avoiding getting sunburnt ☐ D Eating a high fat diet ☐

HT

3 Fill in the missing words to complete the sentences below:

a) A tumour that grows in one place is ....

b) A tumour that spreads to other parts of the body is ....

Infectious Diseases

4 Which of the following are types of pathogen? Tick the four correct options.

A Fungi ☐ B Bacteria ☐

C Tumour ☐ D Organ ☐

E Virus ☐ F Protazoa ☐

Malaria

5 What word is used to describe the type of microorganism that causes malaria? Tick the correct option.

A Fungi ☐ B Bacteria ☐

C Parasite ☐ D Virus ☐

6 Malaria is spread by mosquitoes. What is the term used to describe organisms, like mosquitoes, that spread disease?

....

Keeping Healthy

Defences Against Pathogens

1. The body has several ways to try and protect itself against infection.

   Match defences **A, B, C** and **D** with the sources of infection **1–4** in the table below. Enter the appropriate number in the boxes provided.

	Sources of Infection
1	Dust particles in the air
2	A small cut
3	Bacteria in the soil
4	Bacteria on a cheese sandwich

A Scab ☐ **B** Mucus ☐

C Skin ☐ **D** Stomach Acid ☐

Dealing with Pathogens Inside the Body

2. Which type of body cell is responsible for fighting infection?

   ....

3. Number the statements **1–4** to show the correct sequence of events when a phagocyte encounters pathogens in the body.

   A Phagocytes engulf the pathogens. ☐ **B** The phagocytes digest the pathogens. ☐

   C Pathogens invade the body. ☐ **D** The pathogens are destroyed. ☐

4. Which of the following statements about lymphocytes are correct? Tick the two correct options.

   A Lymphocytes are white blood cells, which produce antibodies. ☐

   B Lymphocytes are pathogens, which attack healthy cells. ☐

   C Lymphocytes use phagocytosis. ☐

   D Antibodies make lymphocytes. ☐

   E Antibodies recognise foreign antigens and stick to them. ☐

HT

5. White blood cells make antibodies that are specific to a particular antigen. Explain what this means.

   .... ☐

Natural (Active) Immunity

1 White blood cells can become **sensitised** to a particular pathogen. What does this mean?

Immunisation

2 Which of the following statements about immunisation is most accurate. Tick the correct option.

A Immunisation is the process by which an individual inherits immunity against a disease from their mother.

B Immunisation provides protection against a disease without the person being infected.

C Immunisation is the process of infecting an individual with a disease to create immunity.

HT

3 Circle the correct options in the following sentences:

Sam was vaccinated with a weakened strain of modified rubella.

a) The weakened **pathogen / phagocyte / protection / particle** is unable to infect Sam with the disease, but his immune system will still respond to it.

b) The **activity / acidity / antigens / red blood cells** on the rubella microorganisms cause his **lymph nodes / lymphocytes / legs / vaccine** to make the specific antibodies, which remain in the blood.

c) If he is later infected with rubella his body will make **protection / the vaccine / antibodies / antigens** very quickly, so he will not become ill.

4 The statements below outline some different opinions about vaccination. Which opinions might lead to parents deciding not to vaccinate their children? Tick the two correct options.

A Vaccines protect against disease.

B Some diseases have been wiped out by vaccination.

C Vaccinations are available from your GP.

D Vaccines may have side effects in some people.

E Vaccines are not effective and it is improved sanitation which has led to the decrease in disease.

Keeping Healthy

Passive Immunity

1 Which of the following situations is an example of passive immunity? Tick the correct option.

A Rashid was bitten by a dog with rabies. Rabies can be fatal, and Rashid was not immune to it. Doctors protected Rashid by injecting him with ready-made antibodies against the rabies. ☐

B Harry caught chickenpox off his brother when he was a toddler. A few years later, all of Harry's classmates caught chickenpox but he did not. This was because his white blood cells were sensitised to the pathogens. ☐

C Mandeep was going on holiday to Thailand. Her doctor gave her vaccinations so that she would be immune to typhoid and hepatitis A. ☐

Treating Diseases with Drugs

2 Which type of pathogens are antibiotics **not** effective against? Tick the correct option.

A Bacteria ☐
B Viruses ☐
C Protazoans ☐
D Fungi ☐

HT

3 Which of these statements about MRSA is true? Tick the three correct options.

A It is also called a superbug. ☐
B It is a bacteria which has become resistant to antibiotics. ☐
C It is a virus so can't be treated with antibiotics. ☐
D It can be very dangerous. ☐
E It is fungal. ☐
F It can be treated with most antibiotics. ☐

4 New medicines must be tested before they are used on patients.
Which of the following statements would apply to a good new medicine? Tick the two correct options.

A It is difficult to take ☐
B It is safe ☐
C It is effective ☐
D It is hard to make ☐
E It is expensive ☐
F It has many side effects ☐

Keeping in Touch

The Nervous System and Neurones

1. Which parts of the nervous system belong to the central nervous system? Tick the two correct options.

   A Effectors ☐ **B** Motor neurones ☐
   C Receptors ☐ **D** Brain ☐
   E Sensory neurones ☐ **F** Spinal cord ☐

2. Where in the body are touch receptors found? Tick the correct option.

   A Skin ☐ **B** Eyes ☐
   C Muscles ☐ **D** Glands ☐

3. What is the long, thin part of a neurone called? Tick the correct option.

   A Axle ☐ **B** Axon ☐
   C Dendrite ☐ **D** Synapse ☐

4. Fill in the missing words to complete the following sentences below:

   a) Sensory neurones carry nerve impulses from the .... to the ....

   b) Motor neurones carry nerve impulses from the .... to the ....

   c) Relay neurones make connections inside your .... and ....

HT

5. Fill in the missing words to complete the following sentences:

   Motor neurones are well adapted to their functions.

   a) They have an .... shape so they can make connections across the body.

   b) They have an insulating sheath to .... the nerve impulse.

   c) They have .... endings so that they can act on many muscle fibres.

HT Synapses

6. The gap between neurones is called a synapse. Describe what happens when an electrical impulse reaches a synapse.

   ....

   ....

Keeping in Touch

Voluntary Actions and Reflex Actions

1. Fill in the missing words to complete the following sentences:

A .... action is a fast, automatic response, which happens in order to protect the body. A slower response, which involves a conscious decision being made, is called a .... action. The difference between these two types of response is that the nerve impulse for a .... action bypasses the ....

2. The diagram illustrates the events that occur during a reflex action.
Match statements **A, B, C** and **D** with the labels **1–4** on the diagram.
Enter the appropriate number in the boxes provided.

A Motor neurone ☐

B Receptor ☐

C Sensory neurone ☐

D Effector cells, e.g. muscle ☐

Relay neurone
Spinal cord
Spinal nerve
4
3
2
1
Drawing pin (stimulus)

The Eye

3. The eye is a complicated sense organ.

Match statements **A, B, C** and **D** with the labels **1–4** on the diagram. Enter the appropriate number in the boxes provided.

A Lens ☐

B Retina ☐

C Cornea ☐

D Iris ☐

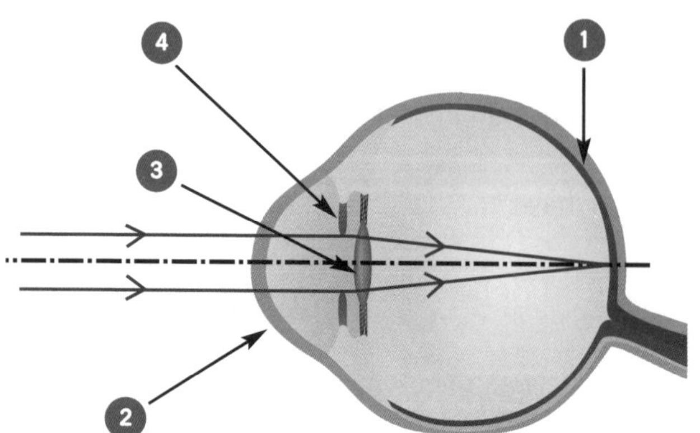

Eye Defects

1 Which of the following statements about red-green colour blindness are true? Tick the two correct options.

- **A** It is an inherited condition. ☐
- **B** It is caused by the lens being misshapen. ☐
- **C** Some of the colour-sensitive cells in the retina are missing. ☐
- **D** It is caused by weak ciliary muscles. ☐
- **E** A person with the condition only sees black and white images. ☐

HT

2 Which of the following would cause a person to have problems with their eyesight? Tick the three correct options.

- **A** Cataracts ☐
- **B** Weak ciliary muscles ☐
- **C** Odd shaped eyeball ☐
- **D** Contact lenses ☐
- **E** Different coloured irises ☐
- **F** Too little sleep ☐

3 **a)** What two defects might cause long sight?

i) ______________________ **ii)** ______________________

b) List three ways in which long sight can be corrected.

i) ______________________ **ii)** ______________________

iii) ______________________

Types of Vision

4 Answer each of the questions below using the word **binocular** or the word **monocular**.

a) What type of vision is best for judging distance and speed? ______________________

b) What type of vision allows an organism to see behind and in front? ______________________

c) What type of vision is commonly found animals at the top of the food chain? ______________________

d) What type of vision is commonly found on animals at the bottom of the food chain? ______________________

e) What type of vision do humans have? ______________________

Drugs and You

Drugs

1 Match effects **A, B, C** and **D**, with the types of drug **1–4** in the table below.
Enter the appropriate number in the boxes provided.

	Category of drug
1	Stimulant
2	Sedative
3	Painkiller
4	Hallucinogen

A Makes you see or hear things that are not there ☐

B Slows down the brain and makes you feel sleepy ☐

C Causes increased brain activity and alertness ☐

D Reduces feelings of pain ☐

HT

2 Fill in the missing words to complete the sentences below:

a) Stimulants .... the amount of transmitter substance released at synapses.

b) Depressants .... the amount of transmitter substance released at synapses.

Drug Classification

3 Illegal drugs are classified into three main categories under the Misuse of Drugs Act. Write **A, B** or **C** alongside each statement to show which category it applies to.

a) Has the heaviest prison sentences and fines for possession. ☐

b) Includes the most dangerous drugs like heroine and cocaine. ☐

c) Carries the lowest penalties for possession, but is still illegal. ☐

d) Includes drugs like cannabis and anabolic steroids. ☐

e) Includes amphetamines like speed and barbiturates. ☐

Addiction and Rehabilitation

4 Addicts' bodies become used to drugs, and they start to need more and more to have the same effect. What is this called? Tick the correct option.

A Addiction ☐ **B** Tolerance ☐

C Withdrawal ☐ **D** Rehabilitation ☐

5 What are two types of withdrawal symptom?

a) .... **b)** ....

Alcohol

1 Alcohol is a commonly used legal drug. What type of drug is alcohol? Tick the correct option.

A Sedative ☐ **B** Analgesic ☐

C Stimulant ☐ **D** Hallucinogen ☐

2 **a)** Which of the following are **not** common short-term effects of drinking too much alcohol? Tick the three correct options.

A Slurred speech ☐ **B** Lack of balance ☐

C Blurred vision ☐ **D** Liver damage ☐

E Brain damage ☐ **F** Damage to unborn child ☐

b) Fill in the missing word to complete the following sentence:

In the long-term, excessive alcohol consumption can damage the liver causing a disease called ....

Tobacco

3 **a)** Despite the risks, many people continue to smoke cigarettes. Tobacco smoke can be very dangerous. Which of these chemicals is found in smoke? Tick the three correct options.

A Tar ☐ **B** Carbon monoxide ☐

C Nicotine ☐ **D** Cilia ☐

E Ketamine ☐

b) Which of these illnesses is **not** directly associated with smoking tobacco? Tick the three correct options.

A High blood pressure ☐ **B** Emphysema ☐

C Cirrhosis ☐ **D** Bronchitis ☐

E Arteriosclerosis ☐ **F** Paranoia ☐

G Pneumonia ☐

HT

4 Which two components of cigarette smoke can cause cancer?

a) .... **b)** ....

Staying in Balance

Homeostasis

1 What is the main purpose of homeostasis? Tick the correct option.

- **A** To control fertility
- **B** To produce a response to external stimuli
- **C** To maintain a constant internal environment
- **D** To provide energy for living cells

2 Which of the following body conditions do **not** need to be kept at a steady level? Tick the two correct boxes.

- **A** Body temperature
- **B** Length of fingernails
- **C** Amount of oxygen in blood
- **D** Amount of carbon dioxide in blood
- **E** Amount of oestrogen in women
- **F** Amount of water in the body
- **G** Blood sugar levels
- **H** Amount of water in the bladder

Temperature Control

3 **a)** Where does the human body get most of its heat from? Tick the three correct options.

- **A** The sun
- **B** Radiators
- **C** Respiration
- **D** Excretion

b) Which of the following happen if the body temperature gets too low? Tick the three correct options.

- **A** Shivering
- **B** Reduction of blood flow to the skin
- **C** Sweating
- **D** Hairs stand on end
- **E** Thirst

c) If the body temperature drops too low it can be very dangerous? What is this condition called? Tick the three correct option.

- **A** Hypoglycaemia
- **B** Hypothermia
- **C** Hormone replacement
- **D** Homeostasis

4 Circle the correct options in the following sentences:

a) When the body is too hot, blood vessels near the surface get **wider / narrower**. This is called **vasodilation / vasoconstriction**. This **reduces / increases** heat loss.

b) When the body is too cold, blood vessels near the surface get **wider / narrower**. This is called **vasodilation / vasoconstriction**. This **reduces / increases** heat loss.

Hormones and Diabetes

1 Select the correct words from the options provided to fill in the spaces and complete the sentences below:

target organs **blood** **endocrine glands** **exocrine glands**

Hormones are made by .... and released into the .... They travel to .... where they have an effect.

2 Which of the following statements about nervous and hormonal control of the body is **not** correct? Tick the two correct options.

A Hormonal responses are faster than nervous ones. ☐

B Hormones are carried in the blood. ☐

C Hormones are chemical messengers. ☐

D Nervous responses last longer than hormonal ones. ☐

E Nerves carry electrical messages. ☐

3 Hormones are important in controlling sexual characteristics.
Which of the following are the main sex hormones? Tick the three correct options.

A Testosterone ☐ **B** Insulin ☐

C Oestrogen ☐ **D** Puberty ☐

E Progesterone ☐

4 **a)** Name two secondary sexual characteristics that develop in girls during puberty.

i) .... **ii)** ....

b) Give two secondary sexual characteristics that develop in boys during puberty.

i) .... **ii)** ....

5 Fill in the missing words to complete the sentences below:

a) Diabetes is caused when the body does not make enough ....

b) Insulin controls the levels of .... in the blood.

HT

c) Insulin helps convert glucose to .... so that it can be stored in the ....

Gene Control

Genetic Information

1 Match statements **A, B, C** and **D**, with the genetic terms **1 - 4** in the table. Enter the appropriate number in the boxes provided.

	Terms
1	Nucleus
2	Chromosomes
3	Genes
4	Bases

A Four chemical building blocks, which hold DNA strands together

B The part of the cell which contains the organisms genetic information

C Long lengths of DNA

D Small pieces of DNA which control the development of a characteristic

HT

2 All the cells in an organism contain the same genes. If this is the case, why are the cells not all identical?

....

3 Fill in the missing words to complete the following sentences:

The order of bases in a .... provides a code for the order in which .... should be assembled by the cell to make proteins. Many proteins produced by cells are ....

Reproduction

4 Fill in the missing numbers to complete the following sentences:

A human body cell contains .... chromosomes. Human sperm and egg cells contain .... chromosomes. When a sperm and egg fuse during fertilisation, they produce a cell containing .... chromosomes.

Variation

1 Although identical twins share the same genetic information, there are usually differences between them that make it easy to distinguish between them. What is the reason for this?

....

2 Which of the following are explanations for genetic variations between individuals? Tick the three correct options.

A Mutations ☐

B Differences between individual gametes ☐

C Developing in different conditions ☐

D Accidents ☐

E Random fusion of egg and sperm at fertilisation ☐

F Medicines ☐

G Different environmental conditions ☐

3 Which of the following variations are due to environmental causes? Tick the three correct options.

A Language spoken ☐

B Scars ☐

C Eye colour ☐

D Attached earlobes ☐

E Hair length ☐

F Tongue rolling ☐

4 Some characteristics are determined by a combination of genetic and environmental factors.

Name one example. ....

Inheritance of Sex

5 What would be the gender of an individual with the following combination of sex chromosomes? Write your answers in the spaces provided.

a) XX ....

b) XY ....

HT

6 When fertilisation takes place, does the sperm or egg control the gender of the offspring?

....

☐

Who Am I?

The Terminology of Inheritance

1 Fill in the missing words to complete the following sentences:

When there are different versions of a gene, one version will be .... and one will be .... Usually, the outcome controlled by the .... version of the gene will occur most frequently.

HT Alleles

2 Match statements **A, B, C** and **D**, with the genetic terms **1 - 4** in the table. Enter the appropriate number in the boxes provided.

	Terms
1	Dominant
2	Recessive
3	Homozygous
4	Heterozygous

A An allele that only controls the development of a characteristic if it is present on both chromosomes in a pair. ☐

B When both chromosomes of a pair contain the same allele of a gene. ☐

C An allele that controls the development of a characteristic, even if it is only present on one chromosome of a pair. ☐

D When the chromosomes of a pair contain different alleles of a gene. ☐

3 What is it called when a characteristic is determined by a single pair of alleles? Tick the correct option.

A Dihybrid Inheritance ☐ **B** Monohybrid Inheritance ☐

C Recessive Inheritance ☐ **D** Homozygous Inheritance ☐

4 A man with blue eyes (bb) and a woman with brown eyes (BB) have a baby. Draw a diagram to show all the possible allele combinations and outcomes for the baby.

Inherited Disease

1 Which of the following are inherited diseases? Tick the three correct options.

A Cystic fibrosis ☐
B Malaria ☐
C Rubella ☐
D Sickle cell anaemia ☐
E HIV ☐
F Red-green colour blindness ☐
G Measles ☐

HT

2 Tom and Ruth have a baby with cystic fibrosis. Neither of them have the disease.

a) Draw a genetic diagram to show how this could happen.

b) What is the chance that their next baby will have cystic fibrosis too?

....

Mutations

3 Fill in the missing words to complete the following sentences:

When a spontaneous change occurs in a gene it is called a .... These changes are usually ....

4 Which of the following can increase the frequency of mutation? Tick the four correct options.

A Ultraviolet light ☐
B Radioactive substances ☐
C X-rays ☐
D Alleles ☐
E Water ☐
F Certain chemicals ☐

Fundamental Chemical Concepts

Atoms

1 **a)** Circle the correct options in the following sentence:

i) Atoms have a **negatively / positively** charged nucleus.

ii) Atoms have **negatively / positively** charged electrons that orbit the nucleus.

b) The nucleus of an atom is made up of protons and neutrons. Which type of particle has no relative charge? ....

Elements and Compounds

2 The table contains four terms relating to how different chemicals can be combined.

Match statements **A, B, C** and **D** with the terms **1-4** in the table. Enter the appropriate number in the boxes provided.

	Term
1	Compound
2	Covalent
3	Mixture
4	Ionic

A A combination of two or more substances, in which each element or compound remains separate. ☐

B A chemical bond formed by atoms sharing electrons. ☐

C A chemical bond formed by atoms losing or gaining electrons. ☐

D A substance formed from two or more elements, which are chemically combined. ☐

3 Which of the following words best describes ammonium nitrate (NH_4NO_3)? Tick the correct option.

A Atom ☐ **B** Element ☐

C Compound ☐ **D** Mixture ☐

Formulae

4 Bleach used by hairdressers contains hydrogen peroxide. The formula for hydrogen peroxide is H_2O_2. Fill in the missing numbers to complete the following sentence:

One molecule of hydrogen peroxide contains .... atoms of hydrogen and .... atoms of oxygen.

Fundamental Chemical Concepts

Formulae (cont.)

1 A scientist discovers a hydrocarbon. Its formula is $CH_3(CH_2)_{50}CH_3$.

a) How many carbon atoms does one molecule of the hydrocarbon contain? ....

b) How many hydrogen atoms does one molecule of the hydrocarbon contain? ....

Displayed Formulae

2 What information does a displayed formula provide? Tick the three correct options.

A The colour of the substance ☐

B The different types of atom in a molecule ☐

C The number of each different type of atom ☐

D The size of each different type of atom ☐

E The covalent bonds between the atoms ☐

F The state of the chemical at room temperature ☐

3 What is the displayed formula for carbon dioxide? Tick the correct option.

A CO_2 ☐ **B** O=C=O ☐

C C + O + O ☐ **D** $C + O_2$ ☐

Equations

4 **a)** Fill in the missing words to complete the following sentences:

In a chemical reaction, the substances that you start with are called .... The new substances formed during the reaction are called ....

b) Equations are used to show what happens during a chemical reaction. Although the atoms are arranged differently, there is always the same number of atoms of each element on both sides of an equation. Why is this?

....

5 Fill in the missing numbers to balance the equations below:

a) $CaCO_3 \rightarrow CaO + CO$ ....

b) $H_2SO_4 + Ca(OH)_2 \rightarrow$ .... $H_2O + CaSO_4$

Cooking

Cooking Food

1 Why is food cooked? Tick the three correct options.

A To kill the microorganisms ▢ B To make it more expensive ▢

C To make it easier to digest ▢ D To make it taste better ▢

2 Cooking food causes a chemical change to take place.
Fill in the missing words to complete the following sentences:

When a chemical change takes place new substances are ...., there may be a .... in mass, and there is often an .... change. The change cannot be easily ....

3 Circle the correct options in the following sentences:

a) Eggs and **vegetables / meat** contain a lot of **protein / fat** molecules.

b) These molecules change shape when they are **neutralised / heated**.

HT

4 Circle the correct options in the following sentence:

When molecules change shape during cooking, it is called **decomposition / denaturing**.

Baking Powder

5 Choose from the words provided to fill in the spaces to complete the following sentences:

carbon dioxide **sodium carbonate** **combustion**

thermal decomposition **sodium hydrogen carbonate** **limewater**

a) The substance used to make cakes rise is called ....

b) When it is heated it breaks down to produce .... which is a solid, water which is a liquid, and .... which is a gas.

6 How do you test for carbon dioxide gas? Tick the correct option.

A A splint will relight ▢ B A splint will squeak / pop ▢

C Limewater will turn cloudy / milky ▢ D Limewater will turn blue ▢

Additives

1 This label displays the nutritional information for a tin of spaghetti.

Cheapsave Value Spaghetti

Ingredients

Water, Sugar, Salt, Maize starch, Onion powder, E160a Paprika, E621 Monosodium glutamate, Flavourings.

Nutrition

Typical Composition	per 100g
Energy	270kJ
Protein	1.6g
Carbohydrate	13g
Fat	0.4g
Fibre	0.5g
Sodium	0.8g

a) Which of the following ingredients has the greatest mass in 100g of spaghetti? Tick the correct option.

A Onion Powder

B Sugar

C Salt

D Starch

b) Which two of the ingredients are food additives?

Emulsifiers

2 Which of the following statements about emulsions and emulsifiers are true? Tick the two correct options.

A Salad dressing is an example of an emulsion.

B Salad dressing is a mixture of alcohol and water.

C Hydrophobic means attracted to water.

D Emulsifiers are molecules that help to keep the oil and water in mayonnaise mixed together.

HT

3 Draw lines between the boxes to match each part of the emulsifier molecule to the correct property.

Hydrophobic end	Polar
Hydrophilic end	Non-polar

Active Packaging

4 Circle the correct options in the following sentence:

a) The quality of food can be improved by **new / active** packaging.

b) The packaging material **hardens / reacts / interacts** to changes to the contents or environment.

c) This can improve the **maturity / safety / colour** of the food.

Smells

Perfumes

1 Members of which group of organic molecules are used to make perfumes? Tick the correct option.

- **A** Alkanes
- **B** Alkenes
- **C** Carboxylic acids
- **D** Esters

2 Fill in the missing words to complete the equation.

Alcohol + ... → .... + Water

Properties of Perfume

3 Which four of the following properties are important to manufacturers when making perfume? Tick the four correct options.

- **A** Evaporates easily
- **B** Does not need to be tested on animals
- **C** Non-toxic
- **D** Does not irritate the skin
- **E** Insoluble in water
- **F** Soluble in water

4 By which method does the smell from perfumes reach your nose? Tick the correct option.

- **A** Expansion
- **B** Transportation
- **C** Diffusion
- **D** Contraction

HT

5 **a)** Explain why perfumes are able to evaporate easily from the skin. Your answer should refer to the forces of attraction between molecules in perfume.

....

....

b) Circle the correct options in the following sentence:

Because perfumes evaporate easily, they are described as **transient / volatile / violent / stable**.

Testing Perfumes

6 Some cosmetics are tested on animals like rabbits and guinea pigs.
Give one argument against testing on animals.

....

Describing Solvents

1 Aftershaves have pleasant smells. Many aftershaves contain alcohol. Which of the following words can be best used to describe how alcohol behaves when it is used in this way? Tick the correct option.

A Solid ☐ **B** Solvent ☐

C Solute ☐ **D** Solution ☐

2 **a)** What is a solute? Tick the correct option.

A A substance that does not dissolve in a liquid. ☐

B A liquid used to dissolve a substance. ☐

C A substance that is dissolved. ☐

D A mixture of chemicals in a liquid state. ☐

b) What does insoluble mean? Tick the correct option.

A A substance that does not dissolve in a liquid. ☐

B A liquid used to dissolve a substance. ☐

C A substance that is dissolved. ☐

D A mixture of chemicals in a liquid state. ☐

Solvents

3 Some potassium nitrate was added to water and the mixture was stirred until all the solid had dissolved. Choose the correct words from the options given to complete the following sentences.

saturated **saturation** **solute** **solution** **soluble** **solvent**

a) The potassium nitrate is a ....

b) The water is a ....

c) Once all the solid has dissolved, the final mixture is called a ....

d) When no more potassium nitrate can dissolve in the water, we say that the solution has become

....

4 Which of these substances is not soluble in water? Tick the correct option.

A Nail varnish ☐ **B** Salt ☐

C Sugar ☐ **D** Felt-tip pen ink ☐

Making Crude Oil Useful

Fossil Fuels

1 How long does it take for a fossil fuel to form? Tick the correct option.

A Ten years ☐ **B** Hundreds of years ☐

C Thousands of years ☐ **D** Millions of years ☐

2 Are fossil fuels **renewable** or **non-renewable**? ....

Crude Oil

3 Circle the correct options in the following sentences:

a) Crude oil is found trapped in **permeable / impermeable** rock.

b) When crude oil is extracted it is a **thin / thick** sticky liquid that is **brown / black** in colour.

4 Circle the correct options in the following sentences:

a) Transporting the oil is a **dangerous / safe** procedure: if the oil spills into the sea, it can have a devastating effect on **wildlife / weather**.

b) Oil spills or slicks **float / dissolve** on the sea surface.

c) They block out the **light / temperature** so plankton, which is at the beginning of the food chain, cannot grow.

Fractional Distillation

5 Choose the correct words from the options given to complete the following sentences:

low **mixtures** **fractions** **different** **high** **elements** **particles** **similar**

a) Crude oil is separated into different .... containing .... hydrocarbons using a fractionating column.

b) These fractions contain many substances with .... boiling points.

c) Fractions with .... boiling points 'exit' from the top of the column.

d) Fractions with .... boiling points 'exit' at the bottom of the column.

6 Fill in the missing words to complete the following sentences:

a) Crude oil is a .... of many different hydrocarbon molecules.

b) A hydrocarbon is a .... that contains only .... and carbon atoms.

Making Crude Oil Useful

Cracking

1 Large hydrocarbon molecules can be broken down through a process called by cracking.

a) What is the purpose of cracking hydrocarbon molecules? Tick the correct option.

A To obtain hydrogen gas

B To obtain shorter hydrocarbon molecules

C To form longer hydrocarbons molecules

D To obtain pure carbon

b) What types of chemicals are obtained from cracking a large chain hydrocarbon?
Tick the two correct options.

A Alkanes

B Alcohols

C Alkenes

D Metals

E Inert Gases

c) Circle the correct options in the following sentence:

Cracking is an example of a **neutralisation / combustion / thermal decomposition / reduction** reaction.

d) Why are the products produced by cracking more desirable than the original large hydrocarbon molecules?

....

Obtaining More Petrol

2 What is the name of the fraction from which petrol is obtained? Tick the correct option.

A Paraffin

B Naphtha

C Liquified petroleum gas (LPG)

D Diesel

HT

3 What problems are there surrounding crude oil and petrol? Tick two correct options.

A Crude oil is a source of confllict.

B Supply of petrol exceeds demand.

C Crude oil is found in many parts of the world.

D Crude oil is a cheap resource because it is made naturally.

Making Crude Oil Useful

HT Forces Between Molecules

1 Circle the correct options in the following sentence:

a) In a hydrocarbon molecule there are **strong / weak** bonds between the atoms.

b) There are **strong / weak** forces of attraction between hydrocarbon molecules.

2 Fill in the missing words to complete the following sentences:

In comparison to short hydrocarbon molecules, longer hydrocarbon molecules have a

.... surface area to come in contact with neighbouring ....

As a result, the forces between longer hydrocarbon molecules are ....

3 Some of the properties of hydrocarbon molecules are shown below.

a) What are the properties of short chain hydrocarbon molecules? Tick the four correct options.

A Runny	☐	**B** Viscous (less runny)	☐
C Easy to ignite	☐	**D** Hard to ignite	☐
E Not used as fuels	☐	**F** Valuable as fuels	☐
G Low boiling points	☐	**H** High boiling points	☐

b) What are the properties of longer chain hydrocarbon molecules? Tick the four correct options.

A Runny	☐	**B** Viscous (less runny)	☐
C Easy to ignite	☐	**D** Hard to ignite	☐
E Not used as fuels	☐	**F** Valuable as fuels	☐
G Low boiling points	☐	**H** High boiling points	☐

c) What property of hydrocarbons allows them to be separated from a mixture by fractional distillation? Tick the correct option.

A Colour	☐	**B** Viscosity	☐
C Smell	☐	**D** Boiling Point	☐

d) Describe the relationship between the boiling point of a hydrocarbon and its viscosity.

....

....

Hydrocarbons

1 **a)** How many bonds does a carbon atom have the ability to form? Tick the correct option.

A 1 ▢ **B** 2 ▢

C 3 ▢ **D** 4 ▢

b) The start of a hydrocarbon's chemical name tells you how many carbon atoms it contains. Number the following prefixes **1-4** to show the number of carbon atoms they represent.

A prop- ▢ **B** eth- ▢

C meth- ▢ **D** but- ▢

HT

2 What type of bonds exist between the hydrogen and carbon atoms in a hydrocarbon?

____

Alkanes and Alkenes

3 Which of the following compounds is an alkene? Tick the correct option.

A Methane ▢ **B** Ethane ▢

C Propane ▢ **D** Butene ▢

4

Diagram 1

```
H     H
 \   /
  C=C
 /   \
H     H
```

Diagram 2

```
H       H
 \      |
  C=C - C - H
 /  |   |
H   H   H
```

Diagram 3

```
    H   H   H
    |   |   |
H - C - C - C - H
    |   |   |
    H   H   H
```

Diagram 4

```
    H   H
    |   |
H - C - C - H
    |   |
    H   H
```

a) Which of the diagrams show alkanes? Tick the two correct options.

A Diagram 1 ▢ **B** Diagram 2 ▢

C Diagram 3 ▢ **D** Diagram 4 ▢

b) The diagrams show the structural formula of four different hydrocarbons.

Match the chemical names **A, B, C** and **D** with the diagrams **1-4**. Enter the appropriate number in the boxes provided.

A Ethane ▢ **B** Ethene ▢

C Propane ▢ **D** Propene ▢

Making Polymers

HT Test for Alkenes

1. Which family of compounds turn bromine water from orange to colourless? Tick the correct option.

   A Alkenes ☐ **B** Alkanes ☐
   C Alcohols ☐ **D** Esters ☐

2. The monomer tetrafluoroethene is unsaturated. What does this mean?

   ....

Polymerisation

3. Fill in the missing words to complete the following sentences:

   The alkenes made by cracking function as .... They can be joined together to produce .... These are ....-chain molecules.

4. **a)** Which group of useful materials is made from polymers? Tick the correct option.

   A Metal ☐ **B** Wood ☐
   C Plastic ☐ **D** Textile ☐

   b) Which monomer is used to produce polythene? Tick the correct option.

   A Ethane ☐ **B** Ethene ☐
   C Propene ☐ **D** Propane ☐

   c) Which polymer is produced from the monomer propene? Tick the correct option.

   A Polyvinylchloride ☐ **B** Polythene ☐
   C Polypropene ☐ **D** Teflon ☐

   d) What conditions are needed to produce the polymer poly(ethene)? Tick the two correct options.

   A Low temperature ☐ **B** High Pressure ☐
   C Presence of catalyst ☐ **D** Presence of oxygen ☐

5. In the space below, write the standard way of displaying a polymer formula.

Designer Polymers

Polymers

1 Polymers have many properties that make them useful.

a) The list below shows eight different properties of polymers (plastics). Which of these properties would be useful when making plastic bags? Tick the four correct options.

A Electrical insulator ☐ **B** Easy to mould into shape ☐
C Waterproof ☐ **D** Flexible ☐
E Rigid ☐ **F** Lightweight ☐

b) Which of the following polymers is commonly used to make plastic bags? Tick the correct option.

A Polythene ☐ **B** Polystyrene ☐
C Nylon ☐ **D** Polyester ☐

c) Which of the following polymers is used to make climbing ropes? Tick the correct option.

A Polythene ☐ **B** Polystyrene ☐
C Nylon ☐ **D** Polyester ☐

d) Which of the following polymers is used to make insulation and packaging (when expanded as foam)? Tick the correct option.

A Polythene ☐ **B** Polystyrene ☐
C Nylon ☐ **D** Polyester ☐

e) Which of the following polymers is used to make clothing and bottles? Tick the correct option.

A Polythene ☐ **B** Polystyrene ☐
C Nylon ☐ **D** Polyester ☐

HT Structure of Plastics

2

a) Which polymer will have a low melting point? ....

b) Which polymer will be flexible and easily moulded? ....

Designer Polymers

Nylon

1 Which of the following is **not** a property of nylon? Tick the correct option.

A Tough ☐ **B** Blocks UV light ☐

C Waterproof ☐ **D** Rigid ☐

Gore-Tex®

2 **a)** Circle the correct options in the following sentences:

Gore-Tex® is a **breathable / coloured / electrical / cheap** material made from **ethane / polystyrene / polythene / nylon**. It has all the advantages of nylon, but it is also treated with a material that allows **carbon dioxide / oxygen / water / nitrogen** vapour to escape whilst preventing rain from getting in.

b) Fill in the missing words to complete the following sentences:

Gore-Tex® is used to make outdoor wear, like rain coats and walking boots. This is because it prevents

.... from getting in, but allows .... to get out.

c) What word is used to describe this property of Gore-Tex®? Tick the correct option.

A Semi-permeable ☐ **B** Absorbent ☐

C Breathable ☐ **D** Membranous ☐

HT

3 In Gore-Tex®, the fibres are coated with a membrane of poly(tetrafluoroethene) (PTFE) or polyurethane. What is the purpose of this coating?

....

Disposal of Plastics

5 The list below suggests some problems caused by the disposal of plastics. Which apply to the disposal of plastics by burning? Tick the two correct options.

A Plastics are non-biodegradable. ☐ **B** It contributes to the greenhouse effect. ☐

C It is very time-consuming. ☐ **D** It produces toxic fumes. ☐

E It uses valuable land. ☐

4 Circle the correct options in the following sentence:

Many plastics are **non-biodegradable / biodegradable / perishable**. This means that they do not **burn / dissolve / decompose**.

Using Carbon Fuels

Combustion

1 When fuels are burned they mainly release thermal energy.

Is this statement **true** or **false**? ....

2 **a)** Hydrocarbons are commonly used as fuels. Which atoms are found in a hydrocarbon molecule? Tick the correct option.

A Hydrogen and oxygen ☐
B Hydrogen and carbon ☐
C Hydrogen, carbon and oxygen ☐
D Carbon and oxygen ☐

b) Complete combustion occurs when a fuel burns and there is enough oxygen. What are the products of complete combustion of a hydrocarbon fuel? Tick the correct option.

A Carbon and water ☐
B Carbon monoxide and water ☐
C Carbon dioxide and water ☐
D Water vapour ☐

3 **a)** Fill in the missing words to complete the following equation for complete combustion of methane:

Methane + ... ➜ .... + Water

HT

b) Fill in the missing numbers to balance the symbol equation for the complete combustion of methane:

CH_4 + ...O_2 ➜ CO_2 + ...H_2O

Incomplete Combustion

4 What is the name of the poisonous gas produced during incomplete combustion? Tick the correct option.

A Carbon dioxide ☐
B Carbon monoxide ☐
C Soot ☐
D Water vapour ☐

5 Which of the following statements about the flame of a Bunsen burner are incorrect? Tick the correct options.

A A blue Bunsen flame transfers more energy than a yellow flame. ☐
B A blue Bunsen flame transfers less energy than a yellow flame. ☐
C A yellow flame produces lots of soot. ☐
D A blue flame on a Bunsen burner shows complete combustion. ☐
E A yellow flame on a Bunsen burner shows complete combustion. ☐

Revision Guide Reference: Page 40

Using Carbon Fuels

Detecting the Products of Combustion

1 The diagram shows the set-up used to detect the products of combustion.

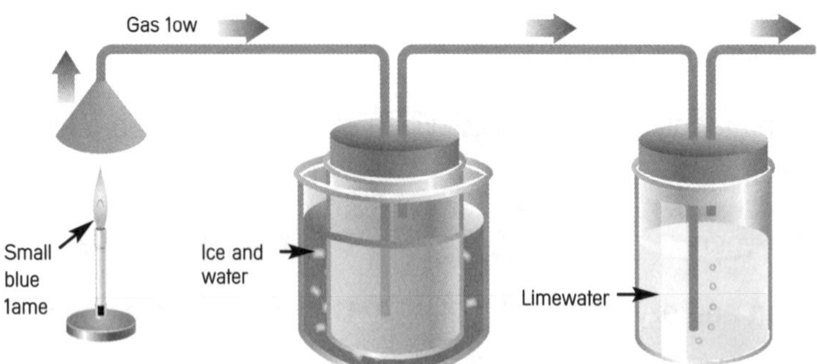

a) What type of fuel is being combusted in this experiment? Tick the correct option.

A Gas ☐ **B** Oil ☐ **C** Coal ☐ **D** Wood ☐

b) Which product of combustion is collected in the beaker in the ice bath? ....

c) Which product of combustion does the limewater in the second beaker test for? ....

d) What will happen to the limewater if this product is present? ....

Choosing a Fuel

2 **a)** What is the energy value of a fuel? Tick the correct option.

A The cost of the fuel per gram ☐

B The amount of energy released per gram ☐

C The amount of energy used in obtaining the fuel ☐

D The cost of obtaining the fuel ☐

3 Below is a list of factors which would influence the fuel chosen to do a particular job.

A energy value

B ease of storage

C cost

D toxicity

E pollution

F ease of use

a) Which factors would be your main concern if you were looking for the fuel that had the least impact on the environment? Write down the letters of two factors. ....

b) Which factors would be your main concern if you wanted to get the highest amount of energy for the least amount of money? Write down the letters of two factors. ....

Exothermic and Endothermic Reactions

1 **a)** What unit is used to measure heat energy? ....

b) What unit is used to measure temperature? ....

2 Circle the correct options in the following sentences:

a) In **calorific / endothermic / exothermic** reactions, energy, usually in the form of **gravity / sound / light / heat**, is given out to the surroundings. This is often shown by **a decrease / an increase** in temperature.

b) In **calorific / endothermic / exothermic** reactions, energy, usually in the form of **gravity / sound / light / heat**, is taken in from the surroundings. This is often shown by **a decrease / an increase** in temperature.

Comparing Fuels

3 What is the name of the technique used to measure the amount of energy released in a chemical reaction? Tick the correct option.

A Titration ☐

B Calorimetry ☐

C Distillation ☐

D Filtration ☐

4 The equipment alongside can be used to compare yield.

a) How would you ensure a fair test? Tick the four correct options.

A Use the same volume of water each time. ☐

B Use a different thermometer each time. ☐

C Only use the draught shield for alternate experiments. ☐

D Use the same calorimeter. ☐

E Use the same mass of fuel. ☐

F Vary the distance between the burner and calorimeter. ☐

G Burn the fuel for the same length of time. ☐

b) Write down the calculation you would use to find the change in temperature of the water.

....

HT Calculating Energy Changes

1 This table shows the results from a calorimetry experiment. In each case...

- 2g of fuel were burnt
- 15cm^3 of water was used (assume 1cm^3 of water has a mass of 1g).

Fuel Used	Water Temperature at Start °C	Water Temperature at End °C
ethanol	20	71
butanol	19	85

Use the information above to answer the following questions:

a) What was the total rise in water temperature when 2g of butanol was burned? Tick the correct option.

A 19°C ▢ **B** 85°C ▢

C 66°C ▢ **D** 51°C ▢

b) What was the mass of water used in the experiment? Tick the correct option.

A 1g ▢ **B** 2g ▢

C 15g ▢ **D** 30g ▢

c) If the speci×c heat capacity of water is 4.2J/g/°C, what was the total amount of energy supplied by the butanol. Tick the correct option.

A 990 joules ▢ **B** 4158 joules ▢

C 3213 joules ▢ **D** 1500 joules ▢

d) What was the energy transfer per gram of butanol? Tick the correct option.

A 2079J/g ▢ **B** 750J/g ▢

C 1607J/g ▢ **D** 1500J/g ▢

Breaking and Making Bonds

2 Circle the correct options in the following sentences:

During chemical reactions, **old bonds / reactions / chemicals / compounds** are broken and new bonds are **broken / required / formed / released**. Energy is **broken / required / formed / released** to break bonds and is **broken / required / formed / released** when new bonds are formed.

Heating Houses

Temperature

1 Choose the correct words from the options given to complete the following sentences:

energy **radiation** **temperature** **heat** **degrees** **Celsius** **therms** **Fahrenheit** **joules**

a) .... is a measure of how hot something is.

b) It's usually measured in ....

c) A hot body transfers .... to a colder body.

d) This is measured in ....

HT

2 a) Fill in the missing words to complete the following sentence:

A thermogram is an .... in which .... is represented by a range of colours.

b) The following colours appear in a thermogram. Write **hot** or **cold** alongside each one as appropriate.

i) Blue ....

ii) Orange ....

Temperature Change

3 The temperature of a cup of tea is measured for ten minutes. The graph shows how the temperature of the tea changed during this time.

Temp °C

80, 60, 40, 20, 0

1 2 3 4 5 6 7 8 9 10

Time min

a) Why didn't the temperature of the tea fall below 20°C? Tick the correct option.

A It stopped giving out heat when it reached the temperature of the room. ☐

B The normal temperature of the tea is 20°C. ☐

C The thermometer wouldn't go below 20°C. ☐

b) Which of the following changes would make the tea cool down more quickly? Tick the correct options.

A Putting a lid on the cup. ☐

B Putting the cup in a warmer room. ☐

C Wrapping a cloth around the cup. ☐

D Standing the cup in cold water. ☐

Measuring Heat Energy

1 The graph shows how the temperature of water in a kettle changes with time. The water temperature rises until it reaches 100°C.

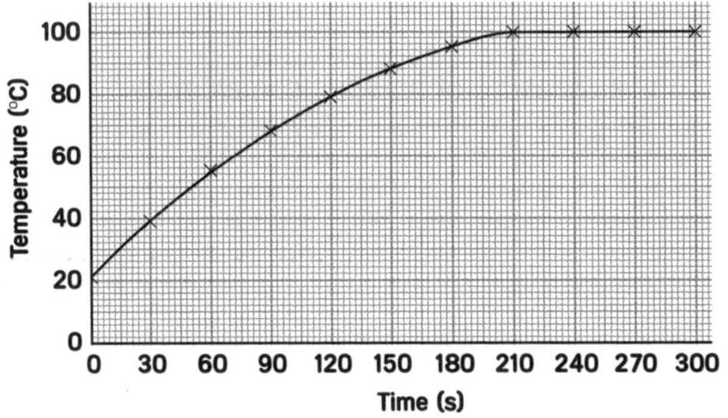

The kettle transferred 1100 joules of energy to the water each second. Use the graph to calculate the total amount of energy transferred to raise the temperature of the water from 20°C to 80°C.

....

Specific Heat Capacity

2 Fill in the missing words to complete the following sentence:

Specific heat capacity is the energy needed to raise the temperature of .... of material by ....

HT

3 In an experiment to measure the specific heat capacity of aluminium, a small electrical heater is used to heat a metal block of mass 0.7kg. Every second 100J of energy passes into the block. A thermometer records the temperature of the block. After one minute the temperature of the block has risen by 9.5°C.

a) How much energy is transferred to the block by the heater in two minutes? Tick the correct option.

A 200J ☐ **B** 180J ☐ **C** 12 000J ☐ **D** 120J ☐

b) Which of these values is the best estimate of the specific heat capacity of aluminium? Tick the correct option.

A 1.5 joules per kilogram per degree Celsius ☐

B 700 joules per kilogram per degree Celsius ☐

C 400 000 joules per kilogram per degree Celsius ☐

D 900 joules per kilogram per degree Celsius ☐

Heating Houses

Melting and Boiling

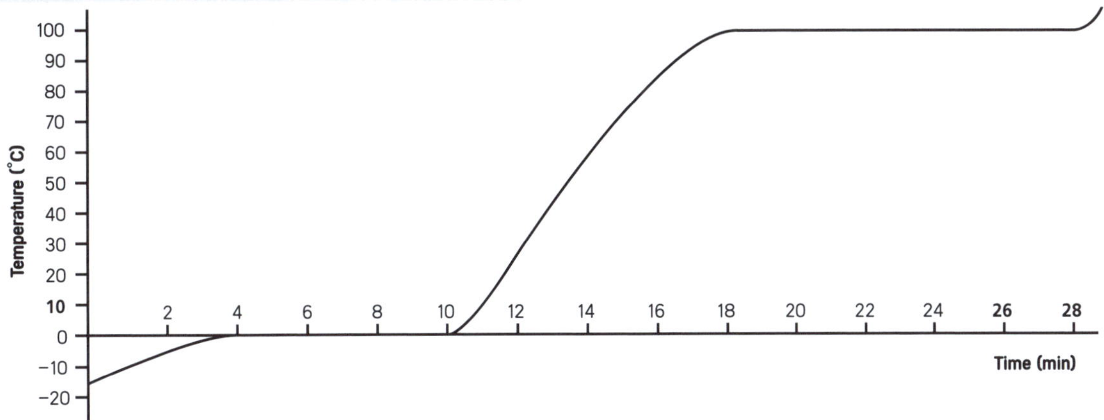

The graph shows how the temperature of a block of ice changes when it is warmed.

a) What is the temperature of the ice, in °C, at the beginning of the experiment? ....

b) At what temperature in °C does the ice melt? ....

c) How many minutes does it take the ice to melt? ....

d) After how many minutes does the water begin to boil? ....

HT

2 Fill in the missing words to complete the following sentences:

a) When a substance melts there is no .... change.

b) The .... supplied to the substance is used to break .... between the .... of the substance.

Specific Latent Heat

3 Which of the following statements about specific latent heat are true? Tick the three correct options.

A The specific latent heat is the amount of energy needed to melt or boil 1 kg of a material. ☐

B The specific latent heat is the amount of energy needed to raise the temperature of 1 kg of a material by 1 °C ☐

C The specific latent heat depends on the material being heated. ☐

D The specific latent heat depends on the state of the material. ☐

Conductors and Insulators

1 Which one of the following is an example of a good conductor? Tick the correct option.

A Plastic ◯ **B** Wood ◯

C Copper ◯ **D** Wool ◯

2 Fill in the missing words to complete the following sentences:

a) A saucepan has a metal base and a wooden handle. This is because the metal is a good .... of ...., so the food heats quickly.

b) The wood of the handle is a good .... and so prevents .... transfer from the saucepan to your hand.

Reducing Heat Losses in the Home

3 The diagram shows how energy is lost from an uninsulated house.

Choose the correct words from the options given to complete the following sentences.

radiators cavity windows insulation
convection conduction walls foil

a) Energy loss through the roof can be reduced by putting .... in the loft.

b) Energy loss through the walls can be reduced by .... wall insulation.

c) Double glazing reduces energy loss through the ....

d) Carpets reduce energy loss by .... through the floor.

4 Describe what is meant by payback time?

....

....

Keeping Homes Warm

HT Saving Energy in the Home

1 Choose the correct words from the options given to complete the following sentences:

radiation	**infrared**	**air**	**solar panels**
water	**cavity wall**	**convection**	**fibreglass**

a) Energy loss through the roof of a house can be reduced by .... This acts as an insulator because it traps ....

b) .... insulation traps air in the foam, reducing energy loss by ....

c) Reflective foil on the walls behind radiators reflects .... radiation back into the room.

2 Explain how double glazing helps reduce energy loss.

....

....

Energy Efficiency

3 Each second 60J of energy is transferred to a lamp. It releases 9J of light energy per second.

a) What is the efficiency of the lamp? Tick the correct option.

A 51% ☐ **B** 25% ☐

C 67% ☐ **D** 15% ☐

b) How many joules of energy are wasted by the light each second? ....

4 Energy saving lamps are now available. A 12W energy saving lamp gives the same light as a 60W conventional lamp. It gives 9J of light energy output for each 12J of electrical energy input.

What is the efficiency of the energy saving lamp? Tick the correct option.

A 30% ☐ **B** 75% ☐

C 13% ☐ **D** 50% ☐

How Insulation Works

Transfer of Heat Energy

1 Hot tea in a cup warms the air above the cup and the air rises. This transfers energy from the tea to its surroundings. What is the name for this type of energy transfer? Tick the correct option.

A Conduction ☐ **B** Convection ☐

C Connection ☐ **D** Radiation ☐

2 Fill in the missing word to complete the following sentence:

Energy is transferred between the Sun and the Earth by ....

HT Conduction

3 Fill in the missing words to complete the following sentences and explain how heat energy is conducted along a metal bar when one end is heated:

a) All particles are .... but those at the hot end of the bar are .... more violently than those at the cold end.

b) The particles at the .... end of the bar jostle against neighbouring particles and transfer ....

c) This makes the neighbouring particles .... more violently. These particles now transfer .... to those further along the bar until the whole bar becomes hotter.

HT Convection

4 Circle the correct options in the following sentence:

Rooms at the top of a building will have a **lower / higher** temperature than those lower down. Warm air will **rise / sink** when allowed to flow and cooler air will **rise / sink** to replace it.

HT Radiation

5 Circle the correct options in the following sentences:

a) Radiation is the transfer of heat energy by **particles / waves**.

b) Dark matt surfaces emit **more / less** radiation than light shiny surfaces.

Cooking with Waves

The Electromagnetic Spectrum

1 The diagram below represents the electromagnetic spectrum, with long wavelengths at one end and short wavelengths at the other.

Use the appropriate letter to mark the approximate position of the following radiations on the diagram:

A Ultraviolet waves **B** Infrared waves **C** Radio waves **D** X-rays

2 Fill in the missing words to complete the following sentences:

a) .... are used to transmit radio and television programmes.

b) .... can give a sunburn and can cause skin cancer.

c) .... are used in medical diagnosis.

Uses of Electromagnetic Radiation

3 What type of radiation does a toaster use to toast bread? Tick the correct option.

A Microwave ☐ **B** Ultraviolet ☐

C Infrared ☐ **D** Visible ☐

4 Decide whether each of the following statements about microwaves is **true** or **false**.

a) They heat food from the middle of the food outward. ....

b) They can be reflected by metal sheets. ....

c) They are emitted from mobile phones. ....

d) They cannot be absorbed. ....

HT Transferring Energy

5 Describe, in terms of particles, how a microwave oven heats food.

....

....

Infrared Signals

1 Which of the following does not use infrared radiation? Tick the correct option.

A Thermograms

B Short distance data link for computers

C Automatic door sensors

D Scans of foetuses in the womb

Relection and Refraction

2 Jack is using a remote control to operate his television. The control works even when it is pointed directly away from the television. Why is this? Tick the correct option.

A Infrared radiation is reflected.

B Infrared radiation is diffracted.

C Infrared radiation is refracted.

D Infrared radiation is focused.

3

Match labels **A, B, C** and **D** with numbers **1-4** on the drawing. Enter the appropriate number in the space provided.

A Air

B Normal line

C Angle of incidence

D Glass

4 Decide whether each of the following statements about optical fibres is **true** or **false**.

a) An optical fibre carries electrical signals at the speed of light. ....

b) The signals are digital because they are a stream of numbers 0-9. ....

c) The signals consist of a stream of pulses. ....

d) Optical fibres use total internal reflection to carry light along the fibre. ....

Infrared Signals and Wireless Signals

Analogue and Digital Signals

1 Circle the correct options in the following sentences:

Infrared signals can be used in television controls. These signals are **analogous / digital**. Analogue signals vary **continuously / randomly** in **amplitude / number** or frequency. The signals in television controls have only two states: on or off.

2 Which of the following statements about digital signals is true? Tick the correct option.

- **A** They are continuous signals. ☐
- **B** They are louder than analogue signals. ☐
- **C** They travel faster than analogue signals. ☐
- **D** They can be processed by computers. ☐

HT

3 Digital signals along fibres can be 'multiplexed'. What advantage is there to multiplexing signals? Tick the correct option.

- **A** It stops interference. ☐
- **B** It allows more information to be transmitted at the same time. ☐
- **C** It speeds up the signal. ☐
- **D** It decreases signal noise. ☐

Wireless Signals

4 Write down three items that use wireless technology.

a) .... **b)** .... **c)** ....

5 Which of the following are possible benefits of wireless technology? Tick the correct option(s).

- **A** No wire connection needed. ☐
- **B** It is portable and convenient. ☐
- **C** It gives a better quality signal. ☐
- **D** It can be available 24 hours a day. ☐

Radio Waves

6 Fill in the missing word to complete the following sentence:

Radio stations operating on similar frequencies can cause ....

Wireless Signals

Microwaves

1 Decide whether the following statements about microwave communication are **true** or **false**.

a) There have been some health concerns about children using mobile phones. ....

b) Microwave communication uses the same wavelength as a microwave oven. ....

c) Increasing the number of transmitting masts does not improve reception. ....

d) Tall transmitting masts improve reception in mobile phones. ....

2 Microwaves are used to transmit information over large distances that are in line of sight. What does this mean?

....

....

HT Transmitting Signals

3 Which three of the following statements are true? Tick the correct options.

A Radio waves are refracted because of the changes in density of the atmosphere. ☐

B Low frequency radio waves give higher quality radio broadcasts than high frequency radio waves. ☐

C Reflection of radio waves by buildings can cause interference. ☐

4 What is the name of the layer of atmosphere that longer wavelength radio waves are reflected by?

....

5 The refraction and diffraction of radiation can affect communications.

a) How does refraction affect waves?

....

b) How does diffraction affect waves?

....

Light

Light

1

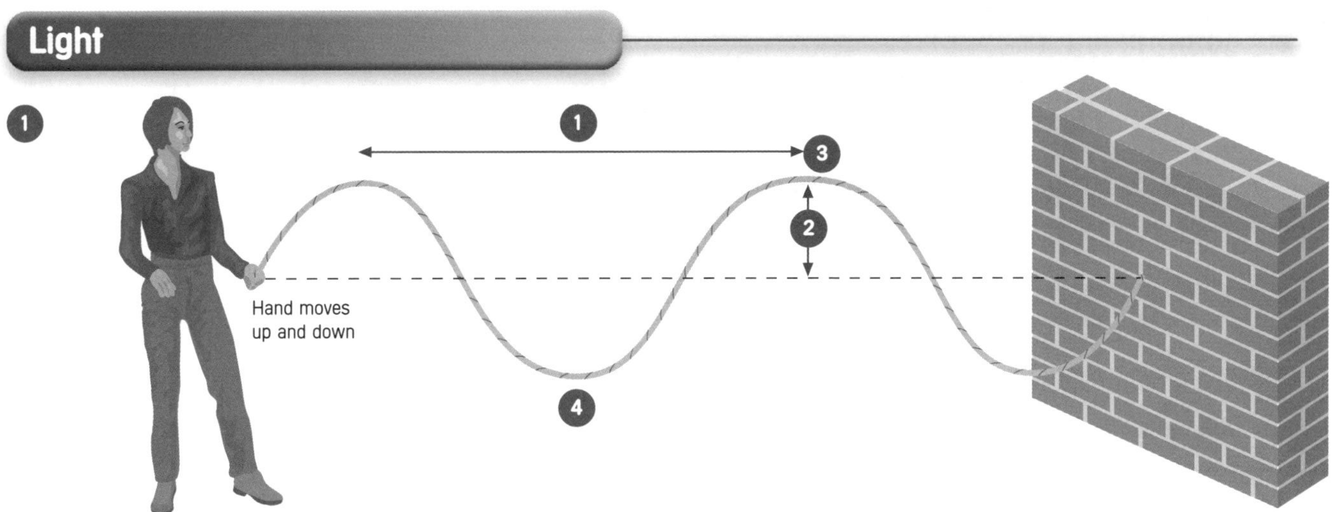

The diagram shows Rosemary demonstrating how light travels by generating a wave with a rope.

a) Match statements **A, B, C** and **D**, with the labels **1–4** on the diagram. Enter the appropriate number in the boxes provided.

A Wavelength ☐ **B** Amplitude ☐

C Trough ☐ **D** Crest ☐

b) In the diagram Rosemary is generating a wave by moving her end of the rope up and down two times every second. What is the frequency of the wave generated by Rosemary? ....

c) The distance between Rosemary and the wall is 3m. What is the wavelength of the wave? Tick the correct option.

A 3m ☐ **B** 6m ☐

C 1.5m ☐ **D** 4.5m ☐

d) What type of wave is light? ....

The Wave Equation

1 This question is about using the equation:

Wave speed = frequency x wavelength
(m/s) (Hz) (m)

When it is struck, a tuning fork produces sound waves of frequency 440Hz. The wavelength of these waves is 0.770m. What is the speed of the sound wave (to the nearest significant figure)? Tick the correct option.

A 570m/s ☐ **B** 0.00175m/s ☐

C 339m.p.h ☐ **D** 339m/s ☐

Communication Signals

1 a) Circle the correct options in the following sentences:

A laser produces a narrow **invisible / intense / weak** beam of light.

b) Give one reason why light is often used to transmit signals in modern technology

....

2 Which of the following is a type of code that can be used to communicate with light? Tick the correct option.

A Number substitution ☐ B Norse ☐

C Horse ☐ D Morse ☐

HT

3 Give one advantage of using radiowaves for communication signals over light.

....

HT Laser

4 Choose the correct words from the options given to complete the following sentences:

laser	**analogue**	**pits**	**coded**
pulses	**invisible**	**reflected**	**digital**

a) In a compact disc .... information can be stored as a sequence of tiny .... in a metal layer.

b) A .... beam is shone onto the spinning disc.

c) The .... pulses of light are then turned into electrical signals.

5 Fill in the missing words to complete the following sentences:

Lasers produce light beams in which all the waves have the same .... The waves in a laser beam are all in .... with each other.

Stable Earth

Earthquakes

1 What is the name of the shock waves that are produced by earthquakes and travel through the Earth?

2 Circle the correct options in the following sentences:

P / S / X waves can travel through solids and liquids. They are **transverse / rectilinear / longitudinal** waves.

3 Which of the following statements about S waves are true? Tick the correct options.

- **A** They are transverse waves.
- **B** They travel faster than P waves.
- **C** They cannot travel through liquids.
- **D** S stands for secondary.

HT Detecting Seismic Waves

4 The illustration shows the structure of the Earth.

Match terms **A, B, C** and **D** below with the labels **1-4** on the diagram. Enter the appropriate number in the boxes provided.

- **A** Mantle
- **B** Crust
- **C** Inner core
- **D** Outer core

1
2
3
4

Global Warming

5 Which of the following does **not** contribute to global warming? Tick the correct option.

- **A** Thinning of the ozone layer
- **B** Increased use of fossil fuels
- **C** Increased CO_2 in the atmosphere
- **D** Deforestation

6 Circle the correct options in the following sentences:

a) Dust from volcanoes reflects radiation from the sun. This causes **warming / cooling**.

b) Dust from factories reflects radiation from cities. This causes **warming / cooling**.

Dangerous Sun

1 What type of radiation causes sunburn? Tick the correct option.

A Infrared ☐ B Ultraviolet ☐

C Gamma ☐ D Visible ☐

2 On a sunny day in England it is advised that you should not spend more than 20 minutes in the sun around midday. Any more time could cause sunburn or longer-lasting damage to the skin. How long could you spend in the sun if you applied a sun block factor 5 cream? Tick the correct option.

A 10 hours ☐ B 5 hours ☐

C 1 hour 20 minutes ☐ D 1 hour 40 minutes ☐

3 Which of the following statements about the effect of ultraviolet radiation on different types of skin are false? Tick the correct options.

A Darker skin absorbs more ultraviolet radiation than lighter skin. ☐

B Darker skin allows less ultraviolet radiation to reach the underlying layers of skin than lighter skin. ☐

C People with darker skin are less at risk from sun damage. ☐

D Light-skinned people are at less risk of getting skin cancer because their skin reflects more of the radiation. ☐

Ozone Depletion

4 What protects the Earth's surface from much of the ultraviolet radiation from the Sun? Tick the correct option.

A Ozone ☐ B The greenhouse gases ☐

C Oxygen ☐ D CFCs ☐

5 Which of the following measures would reduce the thinning of the ozone layer? Tick the correct option.

A Using more solar energy ☐ B Stopping acid rain ☐

C Reducing CFCs in the atmosphere ☐ D Using public transport ☐

6 Many people blame the use of CFCs for the change in the ozone layer.

a) Name two everyday products that use CFCs.

i) .... **ii)** ....

b) What does CFC stand for?

....

Ecology in our School Grounds

Ecological Terms

1 a) What word is used to describe the total number of individuals of the same species that live in an area?

....

b) What is an organism's **habitat**? Tick the correct option.

A The number of individuals in an area ☐

B The number of individuals of one species in an area ☐

C The physical environment where the animal lives ☐

D The physical environment, the conditions, and all the other organisms in an area ☐

Ecosystems

2 a) Which of the following are examples of natural ecosystems? Tick the three correct options.

A Woodland ☐ B Rainforest ☐

C Cattle farm ☐ D Greenhouse ☐

E Lake ☐ F Fish farm ☐

b) Circle the correct option in the following sentence:

Fish farms and greenhouses are examples of **natural / artificial** ecosystems.

HT

3 a) Fill in the missing word to complete the following sentence:

Biodiversity in natural ecosystems is usually .... than in artificial ecosystems.

b) A farmer growing a crop in a greenhouse may use three types of chemicals to maximise his yields. What are these three chemicals called?

i) .... ii) ....

iii) ....

c) Explain how using chemicals to maximise the yield of one crop will effect biodiversity.

....

....

Ecology in our School Grounds

Sampling Methods

1 Toby uses a quadrat to measure the population of dandelions on the school field. The field is 100m² and his quadrat has four equal sides of 0.5m. He throws the quadrat five times and obtains the results shown in the table.

Throws	Number of Dandelions
1	7
2	11
3	8
4	19
5	15

a) What is the area of Toby's quadrat? Tick the correct option.

A $1m^2$ ☐ **B** $0.5m^2$ ☐

C $0.25m^2$ ☐ **D** $2.5m^2$ ☐

HT

b) Why does Toby throw the quadrat five times?

....

....

c) What would you estimate the population of dandelions on the field to be?

....

2 Toby wants to estimate the number of beetles and dandelions in the school grounds.

a) Which of the following should he use? Tick the three correct options.

A Pooter ☐ **B** Key ☐

C Sweepnet ☐ **D** Pitfall trap ☐

E Potometer ☐ **F** Photometer ☐

HT

b) Complete the sentence.

When sampling, you must ensure you take a .... sample. The larger the sample, the more .... the results.

Grouping Organisms

Grouping Animals and Plants

1 Scientists group living organisms together according to their similarities and differences. This process is called classification.

The two main classification groupings are plants and animals. What are the groups called at this level of classification?

....

2 **a)** Which three of the following characteristics are shared by all animals? Tick the correct answers.

A They move around. ☐ **B** They have roots. ☐

C They eat food. ☐ **D** They photosynthesise. ☐

E They have a nervous system. ☐ **F** They have cellulose cell walls. ☐

b) Give three characteristics that are shared by all plants.

i) ....

ii) ....

iii) ....

HT Classification Difficulties

3 **a)** Euglena are microscopic green microorganisms. They can photosynthesise and move around.

i) Which quality do Euglena share with animals?

....

ii) Which quality do Euglena share with plants?

....

b) Scientists have recently recognised that there are five different kingdoms. One of these kingdoms is fungi. Previously fungi were classified as plants.

Describe one way in which fungi differ from plants.

....

....

Grouping Organisms

Vertebrates and Invertebrates

1 a) Animals can be divided into two main groups. Name both of these groups.

i) .... ii) ....

b) Which of the following are types of invertebrates? Tick the correct options.

A Annelids ☐ B Molluscs ☐ C Fish ☐

D Reptiles ☐ E Crustaceans ☐ F Arachnids ☐

c) Draw lines between the boxes to match each type of vertebrate to its characteristics.

Type of Vertebrate	Characteristics
Fish	Produces milk to feed young and have fur.
Amphibian	Has dry scales to prevent water loss on land.
Reptile	Has a moist permeable skin to absorb oxygen from water or air.
Mammal	Has wet scales and gills to obtain oxygen from water.

HT

2 Fill in the missing words to complete the sentences below:

The archaeopteryx is an organism that is now .... We know it existed because it has been identified from ...., which show that it had .... like a bird, but no beak, and .... like a reptile. This makes it very hard to classify.

Species

3 a) What name is given to organisms with similar characteristics that are able to breed together to produce fertile offspring?

....

b) Within the group of organisms described above, what word is used to describe the differences between individuals?

....

Grouping Organisms

Similarities and Differences

1 Different species may share a lot of common features. What are the possible reasons for this? Tick the three correct options.

- **A** The similarities are coincidental. ☐
- **B** They are adapted to survive in the same conditions. ☐
- **C** They have interbred. ☐
- **D** They have a common ancestor. ☐
- **E** They are adapted to compete for the same resources. ☐
- **F** Mutations have produced similar traits. ☐

2 The fossil record sometimes provides evidence that two species, which have a lot of different characteristics, actually share a common ancestor. Why might two related species have different characteristics? Tick the correct option.

- **A** They have interbred with other species. ☐
- **B** Variations caused by environmental factors are passed on to offspring. ☐
- **C** They have adapted over time to survive in different conditions. ☐
- **D** Only one of the species has evolved properly. ☐

HT Evolution and Ecology

3 Dolphins and sharks are examples of two different species that have some similarities.

a) What are some of the similarities between dolphins and sharks? Tick the three correct options.

- **A** They both live in water. ☐
- **B** They both have blow holes. ☐
- **C** They are both covered in scales. ☐
- **D** They both breathe using gills. ☐
- **E** They have similar body shapes. ☐
- **F** They both have fins. ☐

b) What are some of the differences between dolphins and sharks? Tick the three correct options.

- **A** Only one covered in fur. ☐
- **B** Only one is a predator. ☐
- **C** Only one breathes air. ☐
- **D** Only one has a dorsal fin. ☐
- **E** Only one suckles its young. ☐
- **F** Only one has a streamlined body shape. ☐

c) What is the main explanation for the differences between dolphins and sharks?

....

The Food Factory

Photosynthesis

$$\text{Carbon dioxide} + \text{Water} \xrightarrow{\text{light}} \text{Glucose} + \text{Oxygen}$$

a) Which process is represented by the equation above?

....

HT

b) Complete the symbol equation for the same process.

.... $+ 6H_2O \longrightarrow C_6H_{12}O_6 + 6O_2$

Using and Storing Food

2 Write **true** or **false** next to each of the following statements about how plants use the glucose produced during photosynthesis.

a) It is converted to starch. ....

b) It is converted to protein. ....

c) It is used immediately to produce energy. ....

d) It is converted to glycogen. ....

e) It is converted to cellulose. ....

f) It is excreted. ....

g) It is digested. ....

3 Glucose in plants can be converted to several different to meet the plants needs.
Draw lines between the boxes to match each product to its correct use.

Product	Use
Starch	Producing cell walls
Cellulose	Storage in seeds
Protein	Storage in cells
Lipid	Growth and repair

The Food Factory

Increasing Photosynthesis

1 A gardener has a small greenhouse. Which of the following actions would help to increase the growth of the plants in the greenhouse? Tick the three correct options.

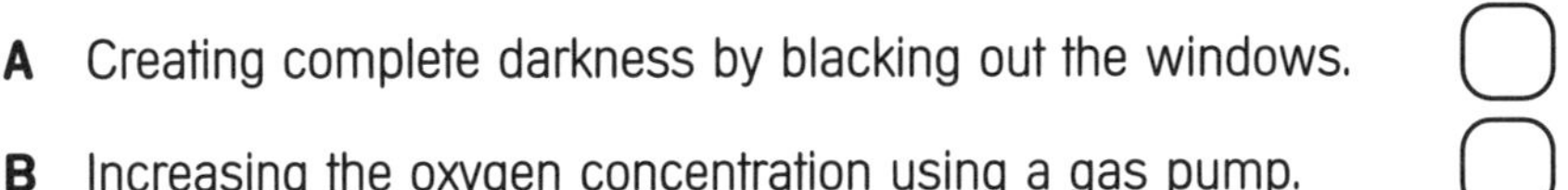

A Creating complete darkness by blacking out the windows. ☐

B Increasing the oxygen concentration using a gas pump. ☐

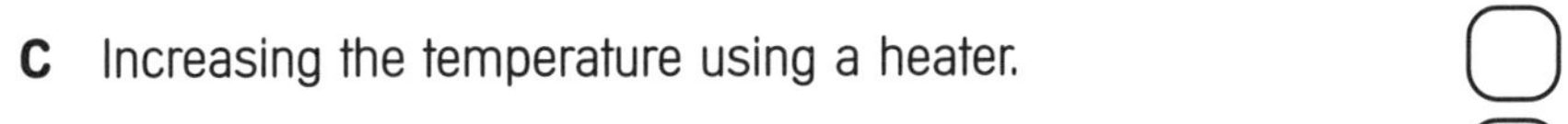

C Increasing the temperature using a heater. ☐

D Increasing the light intensity using a lamp. ☐

E Simulating natural temperatures by opening the windows. ☐

F Increasing the carbon dioxide concentration using chemicals. ☐

HT

2

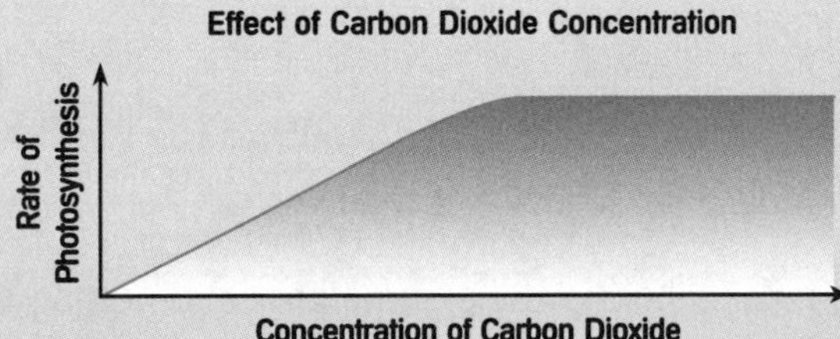

A gardener wanted to make the plants in his greenhouse grow faster, so he added carbon dioxide to the air. The results of this change are shown in the graph.

a) At Point 1 on the graph, what is happening to the rate of photosynthesis as the carbon dioxide level increases? Tick the correct option.

A Nothing ☐ **B** It increases ☐

C It decreases ☐ **D** It stops ☐

b) Explain what is happening at Point 2 on the graph

....

....

Respiration in Plants

3 Why do plants need to respire?

....

Competing for Resources

1 a) Populations of different animals living in the same area compete for which resources? Tick the three correct options.

A Food ☐ B Water ☐

C Shelter ☐ D Light ☐

E Minerals ☐

b) Choose the correct words from the options given to complete the following sentences:

survive **light** **adapted** **resources** **animals**

The better .... competitors will get most of the .... This means they are more likely to .... and reproduce.

HT

c) Fill in the missing words to complete the following sentences:

Organisms living in the same habitat that have the same prey and nesting site are said to occupy the same ....

Predators and Prey

2 The graph shows the numbers of predators and prey in an area over time.

a) What does each arrow show? Tick the correct option.

Match statements **A, B, C** and **D** with the labels **1-4** on the graph. Enter the appropriate number in the boxes provided.

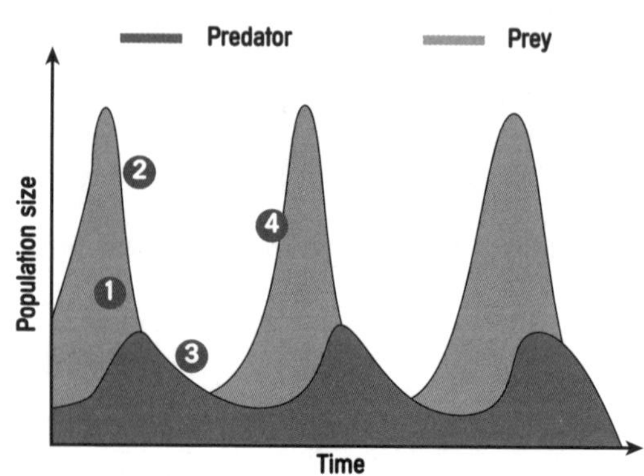

A If predator numbers increase, it causes the numbers of prey to decrease as they are eaten. ☐

B As predator numbers fall the prey numbers can increase again. ☐

C The numbers of predators decrease as there is not enough food for them. ☐

D If the population of prey increases, there is more food for the predator so its numbers increase. ☐

Compete or Die

Parasitic Relationships

1 A parasite is an organism that survives by living off another organism.

a) What word is used to describe the organism that a parasite lives off? ....

b) Name two common parasites.

i) .... **ii)** ....

c) Which of the following statements about parasites is true? Tick the correct options.

- **A** Parasites are dependent on other organisms for survival. ☐
- **B** Parasites benefit the organisms they live off. ☐
- **C** Parasites can cause illness or death in the organism they live off. ☐
- **D** Parasites can survive independently. ☐
- **E** Parasites are normally larger than the organism they live off. ☐

Mutualistic Relationships

2 What is the difference between a mutualistic relationship and a parasitic relationship?

....

....

3 Oxpecker birds have a mutualistic relationship with buffalos.

a) How does the buffalo benefit from the relationship with the oxpecker bird? Tick the two correct options.

- **A** The bird helps to keep the buffalo cool. ☐
- **B** The bird eats flies and parasites. ☐
- **C** The bird grooms the buffalo. ☐
- **D** The bird provides an early warning of predators. ☐
- **E** The bird protects the buffalo's skin from the sun. ☐

b) How does the relationship benefit the oxpecker bird? Tick the correct option.

- **A** The buffalo protects the bird from predators. ☐
- **B** The bird feeds off flies and parasites on the buffalo's skin. ☐
- **C** The buffalo transports the bird from one place to another. ☐
- **D** The buffalo provides an early warning of predators. ☐

Adaptations

1 What are adaptations?

2 **a)** In what type of environment would you expect to find an animal with a large surface area to body ratio, little fat stored under the skin and an ability to tolerate temperature changes without sweating? Tick the correct option.

A Arctic ☐ **B** Rainforest ☐

C Pond ☐ **D** Desert ☐

b) Tigers are native to northern India although adaptation has allowed them to increase their range. Some tigers have longer, thicker fur than the ones in India. Using this fact, sum up how you think their environment differs from their relatives' in India.

Plant Adaptations

3 **a)** Which environment is the cactus adapted to? Tick the correct option.

A Cool and damp ☐ **B** Hot and dry ☐

C Warm and damp ☐ **D** Cold and dry ☐

HT

b) Which of the following features are seen in a cactus? Tick the two correct options.

A Large leaves ☐ **B** Rounded shape ☐

C Brown stem ☐ **D** Thin cuticle ☐

E Long roots ☐

4 **a)** What is the pollen of a wind pollinated plant like? Tick the correct option.

A Small and light ☐ **B** Round and fat ☐

C Sticky ☐ **D** Colourful ☐

b) What is the pollen of an insect pollinated plant like? Tick the correct option.

A Small and light ☐ **B** Smooth ☐

C Sticky ☐ **D** Sweet ☐

Adapt to Fit

Animal Adaptations

1 a) Which of the following adaptations are seen in camels? Tick the four correct options.

- **A** Sandy colour ☐
- **B** Bushy eyelashes and hairy nostrils ☐
- **C** Loses very little water ☐
- **D** Thick layer of fat under the skin ☐
- **E** Fat stored in hump ☐
- **F** Water stored in hump ☐
- **G** Small feet ☐

b) Suggest two ways in which the polar bear is adapted to living in the Arctic.

i) ....

ii) ....

2 Below are examples of how animals have adapted to survive in four extreme environments.

Draw lines between the boxes to match each type of environment to the appropriate adaptation.

Adaptation	Environment
Animals that can resist pressures which would crush you.	Rainforest
Animals that have no eyes but they have increased sensory receptors in other parts of the body.	High mountain
Mammals that have a tail which can grasp hold of things.	Deep ocean
Animals that have large lungs and blood cells which can cope with low levels of oxygen.	Cave

Predator and Prey Adaptations

3 a) Which of the following adaptations are found in many predators? Tick the four correct options.

- **A** Fast over short distances ☐
- **B** Camouflage ☐
- **C** Sharp teeth ☐
- **D** Eyes at the side of the head ☐
- **E** Poisonous sting ☐
- **F** Sharp claws ☐

b) Give two typical characteristics of prey.

i) .... **ii)** ....

Survival of the Fittest

Fossils

1 Occasionally remains of plants or animals from thousands of years ago are found fossilised in rock.

a) Which of the following structures are least likely to form fossils? Tick the correct option.

A Shells ☐ **B** Bones ☐

C Leaves ☐ **D** Brains ☐

b) Which of the following environments could provide suitable conditions for preserving soft tissues? Tick the three correct options.

A A lake ☐ **B** A forest ☐

C A glacier ☐ **D** A tar pits ☐

E A meadow ☐ **F** A peat bog ☐

The Fossil Record

2 **a)** Which of the following are plausible explanations for why the fossil record is incomplete? Tick the four correct options.

A Some body parts may not have been fossilised. ☐

B There is too much water in the air. ☐

C The conditions have to be just right for fossilisation. ☐

D Some fossils might have been lost, particularly if geology changed. ☐

E They weren't all put in the correct place. ☐

F Many fossils are still to be discovered. ☐

3 **a)** The table shows a variety of ways that fossils can form.

Match the types of fossils **A, B, C** and **D**, with the methods of formation **1–4** in the table. Enter the appropriate number in the boxes provided.

	Method of Formation
1	Petrification
2	Preserved in the resin of a tree
3	Preserved in freezing conditions
4	Formed from the hard part of the animal

A 20 million year old insect ☐ **B** Skeleton of tyrannosaurus rex ☐

C 170 million year old tree ☐ **D** A complete woolly mammoth ☐

Survival of the Fittest

The Theory of Evolution

1 As the environment changes, species must also change if they are going to survive.

Choose the correct words from the options given to complete the following sentences:

selection **adapted** **slow** **genes** **evolved**

a) Evolution suggests that all living things have .... from simple life forms developed billions of years ago. The process is .... and continual.

b) Evolution enables organisms to become better .... to their environment.

c) Adaptations are controlled by .... and can therefore be passed on to offspring.

d) Darwin believed that evolution happened by natural ....

2 What might happen to a species which is not well adapted to the environment? Tick the correct option.

A It might become ill ☐
B It might grow bigger ☐
C It might become extinct ☐
D It might have to move ☐

Examples of Natural Selection

3 Most peppered moths are pale and speckled so they are camouflaged on silver birch trees. Until the industrial revolution, dark moths were very rare. During the industrial revolution, the number of dark moths increased around urban areas.

a) Why were dark moths rare before the industrial revolution? Tick the correct option.

A They were bigger ☐
B They had more diseases ☐
C They were more likely to be eaten ☐
D They were mutants ☐

b) Explain why dark moths became more common during the industrial revolution.

....

....

4 Circle the correct options in the following sentence:

Vaccinations / Mutations result in some bacteria becoming resistant to **antibiotics / antibodies**.

This resistance means that they are more likely to **survive / be killed by** penicillin than non-resistant bacteria.

As a result, they **reproduce / die out** and **increase / decrease** in numbers.

Survival of The Fittest

HT Lamarck's Theory

1. Lamarck attempted to explain evolution as the inheritance of acquired characteristics. Why was his theory rejected? Tick the correct option.

   A Organisms do not change. ☐
   B Offspring do not differ from their parents. ☐
   C There was no evidence that acquired characteristics could be passed on to offspring. ☐
   D Lamarck was wrong. ☐

Evolution by Natural Selection

2. What are some of the observations that underpin Darwin's theory of evolution? Tick the three correct options.

   A Population numbers are constantly increasing. ☐
   B All organisms in a species show variation. ☐
   C Acquired characteristics are passed on to offspring. ☐
   D There is competition between individuals in a population. ☐
   E Better adapted individuals are more likely to survive. ☐
   F There is no variation in a species. ☐

3. According to Darwin's theory, what happens to species that are unable to compete?

   ....

4. Fill in the missing words to complete the following sentences:

   From his observations, Darwin deduced that there was a struggle for .... and that the best .... organisms survived. These survivors were able to reproduce and pass on their .... to the next generation.

5. What effect can evolution have on groups of the same species that are separated by physical barriers? Tick the correct option.

   A The two groups evolve identically. ☐
   B The two groups will become extinct. ☐
   C The two groups evolve independently. ☐
   D One of the groups will die out. ☐

Population out of Control?

Pollution

1 The graph shows how the human population is changing.

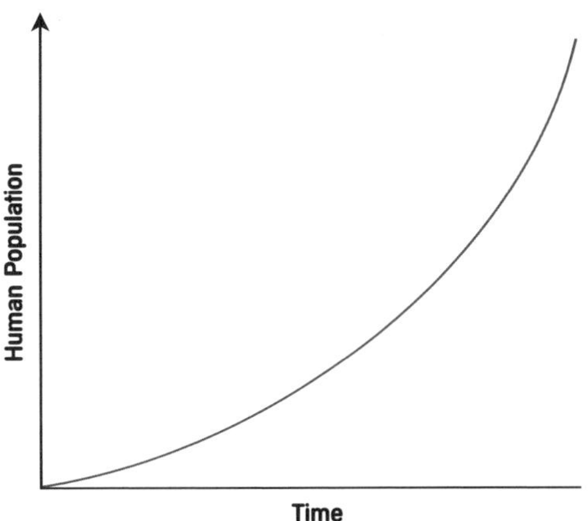

Fill in the missing word to complete the following sentence:

The human population is growing at an accelerating rate. This is known as an .... increase.

3 The growing human population is creating three major issues: increased use of finite resources, increased pollution, and increased competition for basic resources.

a) Which of the following are finite resources? Tick the three correct options.

A Sea water ☐ **B** Coal ☐
C Wind ☐ **D** Minerals ☐
E Wood ☐ **F** Oil ☐

b) Which of the following substances is not contributing to pollution? Tick the correct option.

A Water ☐ **B** Household waste ☐
C Carbon dioxide ☐ **D** Sulfur dioxide ☐

Acid Rain

3 Sulfur dioxide dissolves in water to make acid rain.

a) Where does the sulfur dioxide come from? Tick the correct option.

A It is naturally occurring ☐ **B** The combustion of coal and oil ☐
C Wood smoke ☐ **D** Nuclear power stations ☐

b) Which of the following problems might be caused by acid rain? Tick the three correct options.

A Damage to trees ☐ **B** Statues being eroded ☐ **C** Ozone depletion ☐
D Global warming ☐ **E** Fish dying ☐

Population out of Control?

Ozone Depletion

1 Ozone is an important natural gas.

a) Where is most ozone found? Tick the correct option.

A On the beach

B In exhaust gases

C High up in the atmosphere

D In rainforests

b) What is the function of ozone? Tick the correct option.

A It prevents too many UV rays reaching the earth.

B It contributes to the greenhouse effect.

C It causes pollution.

D It makes acid rain.

Keeping Earth Warm

2 Evidence is being collected to show that the world is warming up. This is known as global warming and is caused by the greenhouse effect.

a) Which of the following gases is contributing to the greenhouse effect? Tick the correct option.

A CFCs

B Ozone

C Carbon dioxide

D Oxygen

b) Which of the following activities release greenhouse gases? Tick three.

A Planting trees

B Melting icecaps

C Burning fossil fuels

D Cutting down forests

E Using aerosols

F Farming cattle

Living Organisms as Indicators

3 Lichen can help to determine pollution levels. Which of the following statements about lichen are true? Tick the two correct options.

A Lichen are sensitive to air pollution.

B Lichen are killed by water pollution.

C Lichen indicate sulfur dioxide levels.

D All lichen respond to pollution in the same way.

Sustainability

Sustainable Development

1 The world's population is growing, increasing the amount of waste produced and the demand for resources. As a result, sustainable development is essential.

a) What is meant by sustainable development?

...

...

...

b) Which of the following methods help to ensure that resources are maintained at suitable levels? Tick the three correct options.

A Intensive farming ☐
B Quotas ☐
C Replenishment ☐
D Classification ☐
E Restocking ☐
F Deforestation ☐

Endangered Species

2 Many species are endangered.

a) What does endangered mean? Tick the correct option.

A They used to exist but don't anymore. ☐
B They are dangerous to humans. ☐
C They are dangerous for the environment. ☐
D They are in danger of becoming extinct. ☐

b) Which of the following organisms are endangered? Tick the two correct options.

A Panda ☐
B Mammoth ☐
C Red squirrel ☐
D Dodo ☐
E Sabre-toothed tiger ☐

c) Which of the following might threaten the survival of a species? Tick the four correct options.

A Climate change ☐
B The introduction of new prey ☐
C Habitat destruction ☐
D Hunting by man ☐
E An increase in resources ☐
F Pollution ☐

HT Conservation Programmes

1 a) What are conservation programmes concerned with? Tick the four correct options.

A Eliminating useless organisms ☐
B Maintaining biodiversity ☐
C Genetic modification ☐
D Stabilising ecosystems ☐
E Studying and identifying plants ☐
F Encouraging intensive farming ☐
G Protecting the culture of indigenous people ☐

b) What are the advantages to society of good conservation? Tick the two correct options.

A More exotic pets are available ☐
B Food supplies are maintained ☐
C New medicines may be developed ☐
D More land is available for development ☐

Whales

2 There are many different whale species but some are now endangered.

a) List four causes of whale deaths.

i) ....

ii) ....

iii) ....

iv) ....

b) One way of conserving whales is to keep them in captivity. What are the problems with this? Tick the two correct options.

A The whales suffer a loss of freedom. ☐
B The whales generate a lot of money. ☐
C The whales are studied. ☐
D The whales do not behave naturally. ☐

HT

c) Name one aspect of whale biology we need to know more about to be able to develop an effective conservation programme.

....

Paints and Pigments

Paint

1 Circle the correct options in the following sentences:

Paint is a special **molecule / mixture / compound / element** of different materials, known as a **colloid / compound / collision / calorimeter**. In paint, fine **aqueous / gas / liquid / solid** particles are well mixed with liquid particles, but they are not **diabetic / displayed / dissolved / disparate**.

2 Match substances **A, B,** and **C** with the descriptions **1-3** in the table. Enter the appropriate number in the boxes provided.

	Description
1	An oil that adheres the pigment to the surface being painted
2	A fine coloured powder that forms a colloid with the binding medium.
3	Has a 'thinning' effect on the thick binding medium, for easier application.

A Pigment ☐ **B** Binding medium ☐ **C** Solvent ☐

HT

3 An oil-based paint dries in two stages. What are they? Tick the two correct options.

A The pigment changes colour. ☐

B The binding medium reacts with oxygen to form a hard layer. ☐

C The solvent and binding medium react to form a gloss finish. ☐

D The solvent evaporates. ☐

E The pigment and binding medium separate. ☐

Special Pigments

4 Fill in the missing word to complete the following sentence:

Pigments that absorb energy and release it as light when it is dark are called ....

Dyes

5 What are the two main advantages associated with manufacturing synthetic dyes?

a) ....

b) ....

Materials from Rocks

1 Which of the following materials come from rocks? Tick the four correct options.

A Plastic ☐ B Aluminium ☐
C Brick ☐ D Wood ☐
E Glass ☐ F Marble ☐

2 Number the following types of stone **1-3**, where 1 is the hardest and 3 is the softest.

A Marble ☐ B Limestone ☐
C Granite ☐

HT

3 a) Why do rocks differ in hardness? Tick the correct option.

A They get harder with age. ☐ B They are formed in different ways. ☐
C Exposure to water softens them. ☐ D Weathering softens them. ☐

Limestone, Cement and Concrete

4 a) Limestone is mainly composed of which chemical compound?

....

b) What is the chemical formula for limestone? ....

c) When limestone is heated it breaks down into two substances.
What is the name given to this type of reaction? Tick the correct option.

A Neutralisation ☐ B Thermal decomposition ☐
C Combustion ☐ D Synthesis ☐

Impact on the Environment

5 Name two ways in which mines and quarries affect their surroundings.

a) ....

b) ....

Does the Earth Move?

Structure of the Earth

1 The drawing shows the structure of the Earth. Match statements **A, B, C** and **D** with the labels **1-4** on the drawing. Enter the appropriate number in the boxes provided.

A Solid inner core ☐ **B** Liquid outer core ☐ **C** Mantle ☐ **D** Thin crust ☐

3
1
2
4

2 What do scientists believe that the Earth's core is made of? Tick the correct option.

A Silicon and carbon ☐ **B** Iron and nickel ☐

C Iron and silicon ☐ **D** Silicon and oxygen ☐

3 What are the waves called that are caused by earthquakes?

....

Movement of the Lithosphere

4 Fill in the missing words to complete the following sentences:

a) The Earth's lithosphere is comprised of the .... and outer ....

The top of the lithosphere is divided into .... plates.

b) Name the two kinds of plates.

i) .... **ii)** ....

c) Name the three ways these plates can move.

i) ....

ii) ....

iii) ....

Does the Earth Move?

HT What Causes Plates to Move?

1 a) What are the currents called that occur within the mantle?

b) What causes these currents?

c) What kind of rock is formed when magma rises and solidifies?

2 Circle the correct options in the following sentences:

Oceanic / continental plates are denser than **oceanic / continental** plates. Therefore a collision between these two types of plates leads to **conduction / subduction / reduction / induction** and partial **remelting / freezing / convection / reduction.**

HT Developing a Theory

3 Give three features that suggest the continents used to be joined together.

a)

b)

c)

4 Fill in the missing words to complete the following sentences:

a) Studies of the ______ of new ______ at plate boundaries were made that support this theory.

b) These studies showed that the plates are moving ______ and the age of rock ______ as you move away from the boundary.

Does the Earth Move?

Volcanoes

1 a) What is magma?

b) Where on the Earth's surface is magma most likely to rise to the surface forming a volcano? Tick the two correct options.

A Where the crust is thin ☐ **B** Where the crust is thick ☐

C In the centre of a tectonic plate ☐ **D** At plate boundaries ☐

E At the two poles ☐ **F** Where there is oceanic crust ☐

2 What type of scientist studies volcanoes?

3 Despite the danger, give one reason why some people chose to live near live volcanoes?

Forming Rock

4 What type of rock is the cone of a volcano made from? Tick the correct option.

A Sedimentary ☐ **B** Igneous ☐

C Metamorphic ☐ **D** Limestone ☐

5 What is the molten rock that erupts from a volcano called? Tick the correct option.

A Magma ☐ **B** Mantle ☐ **C** Lava ☐ **D** Basalt ☐

6 Which of the following are properties of volcanic rock? Tick the two correct options.

A Crystalline structure ☐ **B** Soft ☐

C Granular structure ☐ **D** Hard ☐

7 What determines whether volcanic rock is made of large crystals or small crystals?

HT

8 What are the two different types of lava that can be produced by volcanic eruptions?

a) .... **b)** ....

Metals and Alloys

Copper

1 Fill in the missing words to complete the following sentence:

Copper is extracted from copper ore by .... it with ....

2 Complete the following equation:

Copper oxide + ... → .... **+ Carbon dioxide**

3 Name two benefits of recycling copper.

a) .... **b)** ....

4 Why can recycling copper be difficult?

....

5 Circle the correct options in the following sentence:

Copper used in electrical circuits needs to be **impure / pure / recycled / ignited**.

Electrolysis

6 Fill in the missing words to complete the following sentence:

Electrolysis breaks down a .... made of .... into simpler substances using an ....

7 The diagram shows an electrolysis cell for the purification of copper.

Match statements **A, B, C** and **D** with labels **1-4** on the drawing. Enter the appropriate number in the boxes provided.

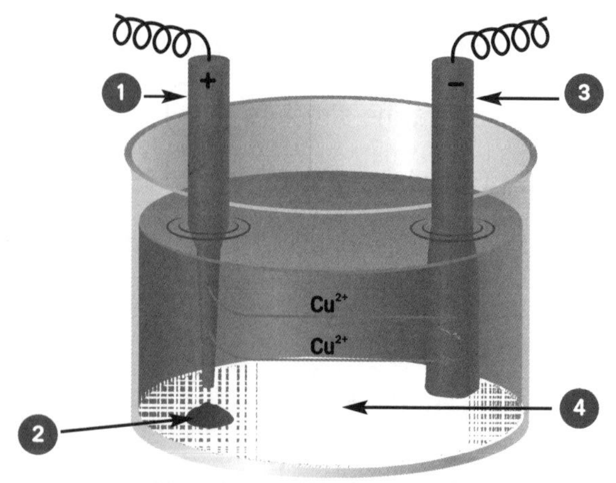

A Pure copper electrode ☐

B Impure copper electrode ☐

C Solution containing copper ions (e.g. copper sulfate solution) ☐

D Impurities ☐

Metals and Alloys

HT Electrolysis

1 In electrolysis, what is the name of the liquid or solution that is used?

....

2 What do the electrodes do?

....

3 a) i) What is the positive electrode called?

....

ii) What is formed at the positive electrode?

....

b) i) What is the negative electrode called?

....

ii) What is formed at the negative electrode?

....

Alloys

4 Fill in the missing words to complete the following sentences:

a) Alloys improve the .... of a metal and make them more useful.

b) Alloys are often .... and .... than the pure metal they are made from.

5 Give two examples of alloys, the metals they are made of, and their uses.

a) Alloy .... Metals ....

Use ....

b) Alloy .... Metals ....

Use ....

Cars for Scrap

Rusting Conditions

1 a) Which two elements are needed to make iron rust? Tick the two correct options.

- **A** Oxygen ☐
- **B** Magnesium ☐
- **C** Aluminium ☐
- **D** Water ☐
- **E** Copper ☐

b) Name a metal that does not rust when exposed to these elements.

....

2 Name two substances that can dramatically increase the rate at which iron rusts.

a) .... **b)** ....

HT

3 **a)** What kind of reaction is rusting?

....

b) Complete the word equation for the rusting of iron:

Iron + ... + ... ⟶ .... iron (III) oxide

Properties of Metals and Alloys

4 Iron can be mixed with the carbon to form the alloy steel. Is this **true** or **false**? ....

5 What two advantages will aluminium have over steel when used to make a car?

a) ....

b) ....

HT

6 How does using aluminium to make a car improve the car's fuel economy?

....

....

Cars for Scrap

Materials in a Car

1 Steel is used to make cars because it has which of the following properties? Tick two correct options.

A Rigid ☐ **B** Malleable ☐

C Strong ☐ **D** Flexible ☐

2 The following substances are used to make cars. Draw lines between the boxes to match each substance to its properties.

Substance	Properties
Nylon fibre seatbelt	Rigid and does not corrode
Glass windscreen	Strong and flexible
Copper wiring in engine	Transparent
Plastic trim	Good electrical conductor

Recycling

3 Most materials in a car can be recycled. Is this **true** or **false**? ....

4 Since 2006, the law has required that 85% of a car must be recycled. What will this increase to in 2015? Tick the correct option.

A 90% ☐ **B** 95% ☐

C 99% ☐ **D** 100% ☐

5 What in a car battery would harm the environment if disposed of without care?

....

6 Fill in the missing words to complete the following sentences:

a) Recycling materials saves resources and reduces .... problems. This protects the environment and saves land being used for ....

b) Recycling plastic helps to conserve .... reserves.

c) Recycling metals requires less .... than extracting them from their ores.

The Atmosphere Today

1 Circle the correct options in the following sentences:

a) The main gas in the atmosphere is **nitrogen / oxygen / carbon dioxide / carbon monoxide.**

b) 21% of the atmosphere is made up of **nitrogen / oxygen / carbon dioxide / carbon monoxide.**

c) Generally, the level of gases in the air is **decreasing / increasing / constant / fluctuating.**

2 a) Which reaction increases the levels of oxygen and decreases the levels of carbon dioxide in the atmosphere? Tick the correct option.

A Combustion ☐ B Neutralisation ☐
C Photosynthesis ☐ D Respiration ☐

b) Which reactions decrease the level of oxygen and increase the level of carbon dioxide in the atmosphere? Tick the two correct options.

A Combustion ☐ B Neutralisation ☐
C Photosynthesis ☐ D Respiration ☐

Changing Levels of Gases

3 a) What two gases did the early atmosphere contain?

i) .... ii) ....

b) Where did these gases come from?

....

c) How were these gases released into the atmosphere?

....

HT

4 Fill in the missing words to complete the following sentence:

As the Earth ...., water vapour in the atmosphere ....

5 Which gas do nitrifying bacteria increase the level of? Tick the correct option.

A Carbon dioxide ☐ B Oxygen ☐
C Nitrogen ☐ D Ammonia ☐

Clean Air

Air Pollution

1 Which of these gases causes acid rain? Tick the two correct options.

A Carbon monoxide ☐

B Carbon dioxide ☐

C Sulfur dioxide ☐

D Nitrogen oxide ☐

2 Name a source of carbon monoxide.

3 Name two problems that are caused by increased pollutants in the atmosphere.

a)

b)

HT Human Influence on the Atmosphere

4 a) How does deforestation affect the amount of carbon dioxide in the atmosphere?

b) Name two other key factors which affect the amount of carbon dioxide in the atmosphere.

i)

ii)

Reducing Pollution

5 a) Circle the correct options in the following sentence:

Catalytic converters change **carbon monoxide / carbon dioxide** into **carbon monoxide / carbon dioxide**.

HT

b) What do catalytic converters contain to cause this reaction?

c) Write out the word equation to show this process.

+ → +

Collision Theory

1 Fill in the missing words to complete the following sentences:

a) For a chemical reaction to take place, the .... particles must collide with each other. They must do this with enough .... to produce a reaction.

b) Reactions occur at .... speeds. The more collisions there are between particles, the .... the reaction.

2 Which of the following will increase the rate of a reaction? Tick the correct three options.

- **A** Increasing the pressure of gases ☐
- **B** Decreasing the temperature ☐
- **C** Increasing the temperature ☐
- **D** Increasing the concentration of liquids ☐
- **E** Decreasing the concentration of liquids ☐

HT

3 Fill in the missing words to complete the following sentence:

The rate of reaction depends on how .... the particles collide and the amount of .... during the collision.

Temperature of the Reactants

4 **a)** What happens when the temperature of a reaction mixture is increased?

....

....

b) What happens when the temperature of a reaction mixture is reduced?

....

....

Faster or Slower (1)

Concentration of the Reactants

1 a) How will particles behave if one or both reactants are in low concentrations?

b) Draw a diagram to illustrate your answer to part a) in the box below.

c) Is the following statement **true** or **false**?

Low concentrations of reactants means there are less successful collisions.

Pressure of a Gas

2 In terms of particles, what happens when the pressure of a gas is increased?

3 If the reactants are under lower pressure there will be more successful collisions.

Is this statement **true** or **false**?

4 The diagrams below show the particles in a gas under high pressure and the particles in a gas under low pressure. Add titles to the diagrams to show which one is which.

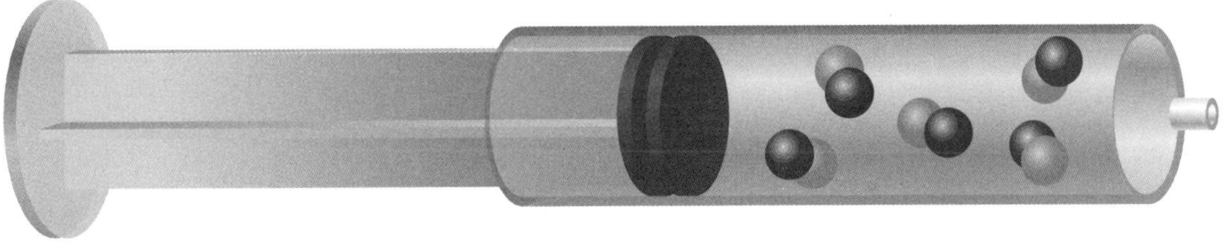

Faster or Slower (1)

Measuring the Rate of Reactions

1 How can the rate of a reaction be measured? Tick the two correct options.

A Measure the temperature ☐

B Look at the equation for the reaction ☐

C Measure how fast the reactants are used ☐

D Reduce the temperature ☐

E Measure how fast the products are formed ☐

2 When does a reaction stop? Tick the correct option.

A When a catalyst is added ☐

B When one of the reactants is used up ☐

C After a couple of days ☐

D When all the products are used up ☐

Analysing the Rate of Reactions

3 The graph shows the rates of reaction for two experiments. Use the graph, along with your knowledge, to answer the following questions.

a) Which line on the graph shows the fastest rate of reaction? ....

b) Give two reasons why this reaction might have taken place at a faster rate.

i) ....

ii) ....

c) The lines on the graph for both experiments level out at the same point. What has happened at this point?

....

....

d) Circle the correct option in the following sentence.

The graph shows the rate of reaction is quickest at the **start / middle / end**.

Faster or Slower (2)

Surface Area of Solid Reactants

1 Circle the correct options in the following sentences.

a) Powdered solids have a **larger / similar / smaller** surface area than a **lump / gas** of the same substance.

b) This means powders have **faster / slower / similar** reactions.

2 Describe why flour must be handled with care in flour mills.

....

....

3 Name two other powdered materials that need to be dealt with cautiously.

a) .... b) ....

Using a Catalyst

4 What does a catalyst do? Tick the correct option.

- **A** It increases the rate of reaction. ☐
- **B** It increases the temperature of the reaction. ☐
- **C** It increases the yield of a reaction. ☐
- **D** It increases the time needed for the reaction. ☐

5 Fill in the missing word to complete the following sentence:

A .... amount of catalyst is needed to increase the rate of reaction.

6 Is the following statement **true** or **false**?

The structure and properties of the catalyst are changed during the reaction. ....

HT

7 When are catalysts most effective?

....

8 Is the following statement is **true** or **false**?

A specific catalyst can be used in many different reactions. ....

Collecting Energy from the Sun

Energy from the Sun

1 Name two types of energy that the Sun transfers to Earth.

a) .... b) ....

2 Circle the correct option in the sentence below.

Energy from the sun is **non-renewable / renewable / finite**.

Photocells

3 Fill in the missing words to complete the sentences:

In photocells, a flat surface made from .... is used to capture solar energy.

The .... is transformed into .... energy.

The type of current produced is called .... current.

4 What does the power output of a photocell depend on?

....

5 Which three of the following options are advantages of using photocells to provide electricity? Tick the three correct options.

A Low maintenance ☐ **B** No polluting waste ☐

C Steady supply of energy ☐ **D** No environmental impact ☐

E No need for fuel ☐ **F** Cheap to install ☐

HT How Photocells Work

6 Choose the correct words from the options given to complete the following sentences. A word may be used more than once:

current **conductors** **heated** **flow** **knocked loose** **reflected** **absorbed**

When light energy is .... by a photocell, some of the silicon's electrons are .... These electrons .... freely. This .... of charge is called a ....

☐

Collecting Energy from the Sun

Other Uses of the Sun's Energy

1 Describe how light from the Sun can be used to heat buildings.

....

2 What does the phrase 'passive solar heating' mean? Tick the correct option.

A A mirror that reflects light but does not move. ☐

B A device that converts the energy of sunlight into electricity which is then used for heating. ☐

C Anything that gets warm in the Sun. ☐

D A device that traps energy from the Sun but does not change it into another form. ☐

HT

3 Why is glass useful in a passive solar heater? Tick the correct option.

A It reflects light. ☐

B It allows all electromagnetic radiation to pass through. ☐

C It allows visible light through but reflects infrared. ☐

D You can see the pipes under the glass. ☐

Wind Turbines

4 Fill in the missing words to complete the following sentences:

Wind turbines depend upon .... in the air produced by the Sun's energy. Wind turbines transfer the .... of the air into ....

HT

5 Which of the following is an advantage of wind turbines? Tick the correct option.

A Wind is a renewable source. ☐

B There is no visual pollution associated with the use of wind turbines. ☐

C They produce power at a steady rate. ☐

D They can be installed anywhere. ☐

Generating Electricity

The Dynamo Effect

1 This question is about the dynamo effect. Use the diagram and your own knowledge to answer the following questions.

Moving the Magnet Towards the Coil of Wire

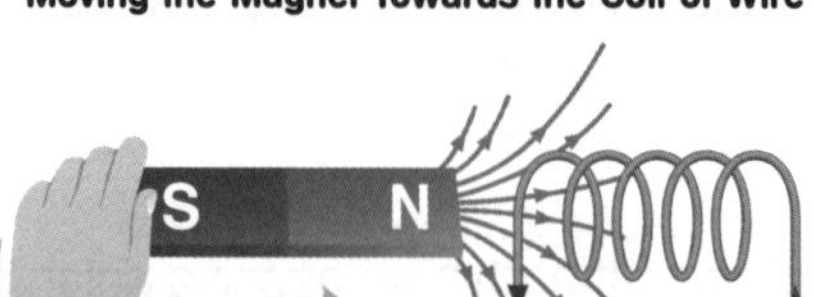

a) What would happen to the reading on the meter if the magnet was moved quickly away from the coil? Tick the correct option.

- **A** The meter would show a bigger deflection to the right. ☐
- **B** The meter would show a small deflection to the left. ☐
- **C** The meter would show a large deflection to the left. ☐
- **D** There would be no change. ☐

b) The magnet is reversed so that the south pole is pointing towards the coil. What will the meter read when the magnet is moved quickly towards the coil? Tick the correct option.

- **A** The meter would show a big deflection to the right. ☐
- **B** The meter would show a small deflection to the left. ☐
- **C** The meter would show a large deflection to the left. ☐
- **D** The meter would show a small deflection to the right. ☐

c) Fill in the missing words to complete the following sentences:

There are three ways in which the current in the coil can be made bigger. The magnet can be moved ....; the number of .... on the coil can be increased; the magnet can be replaced by a .... one.

The AC Generator

2 Choose the correct words from the options given to complete the following sentences:

induced **alternating** **size** **speed** **direction** **direct**

If a coil is rotated between the poles of a magnet, a current is .... in the coil.

The .... of flow changes with every half turn of the coil, so an .... current is produced.

Fuels for Power

Producing Electricity

1 Number the statements **1**-**4** to show the sequence of events for producing electricity.

A The steam drives the turbines, which drives generators. ☐

B The fuel (heat source) is burned to release heat energy. ☐

C The generators produce electricity. ☐

D The heat boils water to produce steam. ☐

2 Explain what **wasted energy** is.

____

____

HT Efficiency of a Power Station

3 A power station uses 400 000J of energy to produce 160 000J of electrical energy.

a) Calculate the waste energy output.

____

b) Calculate the efficiency of the power station.

____

Distributing Electricity

4 Briefly explain why electricity is transmitted through the power lines at a high voltage.

____

HT

5 Fill in the missing words to complete the following sentences:

A step-up transformer increases the ______________ of the transmitted power and reduces the ______________ . This reduces ______________ loss in the power lines.

Power Station Fuels

1 What are the three types of fuel commonly used in power stations?

a) .... b) .... c) ....

2 Choose the correct words from the options given to complete the following sentences:

burned **dissolved** **fermented** **methane** **coal**

Biomass can be .... to produce .... gas. This can be

.... to release heat energy.

3 Which of the following options is not an example of a fuel that is fermented as biomass. Tick the correct option.

A Sheep droppings ☐
B Straw ☐
C Coal ☐
D Fallen trees ☐

Nuclear Power

4 Which of these statements about producing power using uranium is false? Tick the correct option.

A Plutonium can be used to make nuclear bombs. ☐
B Radiation from radioactive waste can cause cancer. ☐
C Plutonium is a waste product from nuclear reactors. ☐
D Uranium is a renewable source. ☐

HT Off-Peak Electricity

5 Electricity boards offer cheaper electricity for seven hours each night. This is called Economy-7. Which of the following statements about Economy-7 is **not** true? Tick the correct option.

A It is used mainly for heating water and storage heaters. ☐
B It is cheaper to persuade people to do more work at night. ☐
C There is less demand for power during the night. ☐
D If more people chose to run washing machines or dishwashers during the night the output of the power stations would not vary so much between day and night. ☐

Fuels for Power

HT Comparing Energy Sources

1 Which of the following are disadvantages of nuclear power? Tick the three correct options.

- **A** High decommissioning costs
- **B** Produces greenhouse gases when uranium is used in a power station
- **C** Low fuel stocks
- **D** Pollution from fuel processing
- **E** High maintenance costs
- **F** Energy supply is not constant

2 Which of the following are advantages of renewable sources? Tick the three correct options.

- **A** Energy supply can be flexible
- **B** Doesn't produce acid rain
- **C** Can be constructed in remote areas
- **D** Doesn't take up much space
- **E** Maintenance costs are low
- **F** Always produce large amounts of energy

3 **a)** Give two advantages of using biomass to produce energy.

i) ....

ii) ....

b) Give two disadvantages of using biomass to produce energy.

i) ....

ii) ....

4 **a)** Name one environmental benefit of using nuclear fuel.

....

b) Name one disadvantage for the environment of using nuclear fuel.

....

c) Do you think the disadvantage outweighs the benefit? Explain your answer.

....

....

Power

1 a) What determines the power of a component or device? Tick the correct option.

A The type of energy transformation ☐ B The rate of energy transfer ☐

C The voltage ☐ D The size of current ☐

b) What unit is used to measure power?

c) What is the standard formula used to calculated power?

2 A kettle draws a current of 11A from the 230V mains. Calculate the power of the kettle.

Kilowatt Hours

3 Which of the following options explains what is happening to the electrical energy of a toaster that has a power rating of 1.1kW? Tick the correct option.

A 1100J of electrical energy changes into heat and light energy every second. ☐

B 1.1J of electrical energy changes into heat and light energy every second. ☐

C 1100J of electrical energy changes into heat and light energy every hour. ☐

D 1100J of electrical energy changes into heat and light energy every minute. ☐

4 Vicki uses a 3kW heater for 2 hours. How much electrical energy in kilowatt hours has been supplied during this time? Tick the correct option.

A 5kW ☐ B 1kW ☐

C 6kW ☐ D 21 600 000kW ☐

5 One unit (kilowatt hour) of electricity costs 8p. How much would it cost if a 3kW heater is used for 3.5 hours?

HT

6 Gary uses an 800W microwave oven to cook his supper. The food takes 15 minutes to heat. How much electrical energy in kilowatt hours has been supplied during this time? Tick the correct option.

A 12 000 kWh ☐ B 12 kWh ☐

C 200 kWh ☐ D 0.2 kWh ☐

Nuclear Radiations

Background Radiation

1. What are the three types of nuclear radiation?

   a) .... b) ....

   c) ....

2. Which of the following statements best describes radiation? Tick the correct option.

   A Radiation is the product of a biological mutation. ☐

   B Radiation is the result of atomic changes in radioactive materials. ☐

   C Radiation is produced my mixing reactants in a chemical reaction. ☐

   D Radiation is only produced by living materials. ☐

3. List three sources of background radiation.

   a) .... b) ....

   c) ....

4. Is background radiation a threat to our health? Explain your answer

   ....

Penetration and Ionisation

5. **a)** What are ions?

   ....

   b) How are ions produced?

   ....

6. **a)** Which type of radiation is completely absorbed by a thin sheet of paper? Tick the correct option.

   A Alpha ☐ **B** Beta ☐

   C Gamma ☐

   b) What type of radiation can only be stopped by an thick sheet off lead? Tick the correct option.

   A Alpha ☐ **B** Beta ☐

   C Gamma ☐

Nuclear Radiations

Uses of Alpha and Beta Radiation

1 Nuclear radiation is used to test the thickness of paper. Which of the following statements correctly describes the type of radiation used in this process and the reason for the choice? Tick the correct option.

- **A** Beta radiation is used because it will pass through thin sheets of paper but not pass through thicker sheets. ☐
- **B** Beta radiation is used because it has a low ionising power. ☐
- **C** Alpha radiation is used because it has a high penetrating power. ☐
- **D** Alpha radiation is used because it is easily stopped. ☐

2 Which of the following is a common use of alpha radiation? Tick the correct option.

- **A** Medical tracer ☐
- **B** Sterilising medical equipment ☐
- **C** Smoke detectors ☐
- **D** Cancer treatment ☐

Uses of Gamma Radiation

3 Which three of the following options are uses of gamma radiation? Tick the correct options.

- **A** Treating cancer ☐
- **B** Sterilising food ☐
- **C** Sterilising medical equipment ☐
- **D** In smoke detectors ☐

Handling Radioactive Materials

4 Explain how higher-level waste is disposed of.

....

....

....

5 Give three safety precautions that need to be taken when handling radioactive material.

a) ....

b) ....

c) ....

Our Magnetic Field

Magnetic Fields

1 The diagram shows the Earth and some field lines.

a) Match statements **A**, **B** and **C**, with the labels **1-3** on the diagram.

A Direction compass needle points ☐

B Magnetic field lines ☐

C Rotation axis ☐

HT

b) What kind of particles can magnetic fields be generated by? ....

Solar Flares and Cosmic Rays

2 What is a solar flare?

....

HT

3 What would be the effect of a solar flare reaching Earth?

....

4 How does the Earth's magnetic field protect its surface from cosmic rays?

A The atmosphere absorbs some of the cosmic rays. ☐

B The cosmic rays create gamma rays. ☐

C The cosmic rays spiral along the magnetic field lines of the Earth to the Poles. ☐

D The magnetic field of the Earth deflects the cosmic rays back into space. ☐

The Earth-Moon System

5 Circle the correct options in the following sentences:

When two planets collided to form the Earth and the Moon, the **iron / metal / carbon** core of each planet **merged / separated** to form the core of the Earth. The Moon was formed out of less **dense / dry / valuable** material.

Exploring our Solar System

The Universe

1 **a)** Stars make up part of the Universe. Which of the following options explain why we can see stars? Tick the correct option.

- **A** They are very close to us. ☐
- **B** We can only see them through a telescope. ☐
- **C** They are very hot and give out light. ☐
- **D** They are very dense so we see light disappearing into them. ☐

b) List the other bodies that occur in the universe.

....

c) What is a galaxy?

....

d) What is a black hole?

....

HT

2 Briefly explain why nothing, not even light, can escape from a black hole.

....

....

Orbits in the Solar System

3 Which force keep planets, comets and satellites in their orbits around the larger body they are orbiting? Tick the correct option.

- **A** Resultant force ☐
- **B** Gravitational force ☐
- **C** Nuclear force ☐
- **D** Radiation ☐

4 Fill in the missing words to complete the following sentences:

The planet that orbits closest to the Sun is .... orbits at the greatest distance from the Sun.

Exploring Our Solar System

Manned Space Travel

1 Describe three problems facing a manned space mission to the planets.

a) ....

....

b) ....

....

c) ....

....

Unmanned Space Travel

2 List four types of information about planets that probes on unmanned space ships can be used to gather.

a) .... b) ....

c) .... d) ....

HT

3 Unmanned spacecraft can withstand conditions that would be lethal to humans.

a) Which of the following options explain why using unmanned spacecraft is useful? Tick the correct options.

A It makes the mission cheaper. ☐

B It makes the spacecraft more robust. ☐

C It means the spacecraft can explore areas that would be impossible for humans to reach. ☐

D It means the spacecraft can go further. ☐

4 What is a light year? Tick the correct option.

A The distance light travels in a month ☐ B The distance light travels in a year ☐

C The distance from the Sun to the Earth ☐ D The distance sound travels in a year ☐

5 Give one disadvantage of using unmanned space ships for space exploration.

....

Threats to Earth

Asteroids

1 Choose the correct words from the options given to complete the following sentences:

Gas **gases** **wind** **rocks** **heat** **dust** **climate**

If an asteroid collided with the Earth there would be many devastating consequences. The impact would form a crater which could begin the ejection of hot .... This might cause fires. The explosion could cause a lot of .... to be thrown into the atmosphere which block out the sunlight and lead to .... change.

HT

2 Why are most asteroids found in a belt between the orbits of Mars and Jupiter? Tick the correct option.

- **A** The strength of Jupiter's gravity has attracted them from far out in the Solar System. ☐
- **B** The strength of Mars' gravity keeps them in place. ☐
- **C** The strength of Jupiter's gravity prevented a planet from forming so the rocks do not join together. ☐
- **D** It is just a coincidence. ☐

Comets

3 What is the characteristic tail of a comet made from? ....

HT

4 Why do comets travel faster nearer the Sun? Tick the correct option.

- **A** They accelerate as they approach the Sun. ☐
- **B** They get more kinetic energy as they approach the Sun. ☐
- **C** They get hotter as they approach the Sun. ☐
- **D** The force of gravity increases as they approach the Sun, accelerating the comets. ☐

Near Earth Objects

5 What instrument is used to measure the possible paths or NEOs?

....

The Big Bang

Big Bang Theory

1 Why is the modern theory of the beginning of the Universe called the 'Big Bang'? Tick the correct option.

- **A** Because the Universe was empty until there was a really big, loud explosion. ☐
- **B** Because the Universe began in a single event and has expanded ever since. ☐
- **C** Because before the explosion the Universe was silent. ☐

2 Which of the following statements is evidence that the Universe is expanding? Tick the correct option.

- **A** Some stars are red. ☐
- **B** Some stars are getting dimmer. ☐
- **C** All the galaxies are moving away from our own galaxy. ☐
- **D** The further away a galaxy is the slower it is moving. ☐

HT

3 Which of the following describes what is meant by the term 'red shift'? Tick the correct option.

- **A** The lengthening of the wavelength of light from a moving object ☐
- **B** The lengthening of the wavelength of light from an object moving away ☐
- **C** The lengthening of the wavelength of light from an object moving nearer ☐
- **D** An object looking red because it is moving ☐

The End of a Star

4 Some stars end up as white dwarves, others as neutron stars and others as black holes. What decides the final state of a star? Tick the correct option.

- **A** The mass of the original star ☐
- **B** The chemical composition of the original star ☐
- **C** The temperature of the star ☐
- **D** The colour of the star ☐

HT The Life of a Star

5 Fill in the missing words to complete the following sentences:

Stars are formed when .... collapse because of .... These form .... Gradually the .... increases and eventually .... fusion takes place and the main sequence star is born. ☐

Notes

Notes

Key

relative atomic mass **atomic symbol** name atomic (proton) number

1	2											3	4	5	6	7	0/8
							1 **H** hydrogen 1										4 **He** helium 2
7 **Li** lithium 3	9 **Be** beryllium 4											11 **B** boron 5	12 **C** carbon 6	14 **N** nitrogen 7	16 **O** oxygen 8	19 **F** fluorine 9	20 **Ne** neon 10
23 **Na** sodium 11	24 **Mg** magnesium 12											27 **Al** aluminium 13	28 **Si** silicon 14	31 **P** phosphorus 15	32 **S** sulfur 16	35.5 **Cl** chlorine 17	40 **Ar** argon 18
39 **K** potassium 19	40 **Ca** calcium 20	45 **Sc** scandium 21	48 **Ti** titanium 22	51 **V** vanadium 23	52 **Cr** chromium 24	55 **Mn** manganese 25	56 **Fe** iron 26	59 **Co** cobalt 27	59 **Ni** nickel 28	63.5 **Cu** copper 29	65 **Zn** zinc 30	70 **Ga** gallium 31	73 **Ge** germanium 32	75 **As** arsenic 33	79 **Se** selenium 34	80 **Br** bromine 35	84 **Kr** krypton 36
85 **Rb** rubidium 37	88 **Sr** strontium 38	89 **Y** yttrium 39	91 **Zr** zirconium 40	93 **Nb** niobium 41	96 **Mo** molybdenum 42	[98] **Tc** technetium 43	101 **Ru** ruthenium 44	103 **Rh** rhodium 45	106 **Pd** palladium 46	108 **Ag** silver 47	112 **Cd** cadmium 48	115 **In** indium 49	119 **Sn** tin 50	122 **Sb** antimony 51	128 **Te** tellurium 52	127 **I** iodine 53	131 **Xe** xenon 54
133 **Cs** caesium 55	137 **Ba** barium 56	139 **La*** lanthanum 57	178 **Hf** hafnium 72	181 **Ta** tantalum 73	184 **W** tungsten 74	186 **Re** rhenium 75	190 **Os** osmium 76	192 **Ir** iridium 77	195 **Pt** platinum 78	197 **Au** gold 79	201 **Hg** mercury 80	204 **Tl** thallium 81	207 **Pb** lead 82	209 **Bi** bismuth 83	[209] **Po** polonium 84	[210] **At** astatine 85	[222] **Rn** radon 86
[223] **Fr** francium 87	[226] **Ra** radium 88	[227] **Ac*** actinium 89	[261] **Rf** rutherfordium 104	[262] **Db** dubnium 105	[266] **Sg** seaborgium 106	[264] **Bh** bohrium 107	[277] **Hs** hassium 108	[268] **Mt** meitnerium 109	[271] **Ds** darmstadtium 110	[272] **Rg** roentgenium 111							

Elements which have atomic numbers 112–116 have been reported but not fully authenticated.

*The lanthanoids (atomic numbers 58–71) and the actinoids (atomic numbers 90–103) have been omitted.

The relative atomic masses of copper and chlorine have not been rounded to the nearest whole number.

→ The lines of elements going across are called **periods**.

↓ The columns of elements going down are called **groups**.